THE FUTURE OF SCIENTIFIC PRACTICE: 'BIO-TECHNO-LOGOS'

I0051274

History and Philosophy of Technoscience

Series Editor: Alfred Nordmann

Titles in this Series

Forthcoming Titles

THE FUTURE OF SCIENTIFIC PRACTICE: 'BIO-TECHNO-LOGOS'

EDITED BY

Marta Bertolaso

Routledge
Taylor & Francis Group

LONDON AND NEW YORK

First published 2015 by Pickering & Chatto (Publishers) Limited

2 Park Square, Milton Park, Abingdon, Oxon OX14 4RN
711 Third Avenue, New York, NY 10017, USA

Routledge is an imprint of the Taylor & Francis Group, an informa business

First issued in paperback 2016

Copyright © Taylor & Francis 2015
Copyright © Marta Bertolaso 2015

To the best of the Publisher's knowledge every effort has been made to contact
relevant copyright holders and to clear any relevant copyright issues.
Any omissions that come to their attention will be remedied in future editions.

All rights reserved. No part of this book may be reprinted or reproduced or
utilised in any form or by any electronic, mechanical, or other means, now known or
hereafter invented, including photocopying and recording, or in any information
storage or retrieval system, without permission in writing from the publishers.

Notice:
Product or corporate names may be trademarks or registered trademarks,
and are used only for identification and explanation without intent to infringe.

BRITISH LIBRARY CATALOGUING IN PUBLICATION DATA

The future of scientific practice: 'bio-techno-logos'. – (History and philosophy
of technoscience)
1. Science – Philosophy.
I. Series II. Bertolaso, Marta, editor.
501-dc23

ISBN-13: 978-1-8489-3562-4 (hbk)
ISBN-13: 978-1-138-70642-2 (pbk)

Typset by Pickering & Chatto (Publishers) Limited

CONTENTS

PREFACE

This volume is the result of informal discussions among the members of the interdisciplinary Bio-Techno-Practice (BTP) think tank based at the University Campus Bio-Medico in Rome working on the dynamics of non-linear systems. The original focus was on cancer and complex diseases and systems biology, but we soon realized that some issues were continuously recurring in our discussions. Questions like: Why is the prefix 'bio' becoming so common in different fields, from informatics to robotics? What kind of methodological and philosophical issues are at stake when we focus on 'biological' dynamics? Is the use of the prefix 'bio-' a mere tribute to a largely media-oriented fashion of identifying biology as the central science of the present or does the reference to biological organization identify a shared need for a change in the way we look at technology? On the other hand, is the explicit reference biology makes to technological tasks and activities recorded by words like biotechnology, bioinformatics, systems and computational biology, which substitute time-honoured synonyms like enzimology, biophysics, physiology, something more than a smart branding initiative?

The same mood that something was missing in our account (and manner of talking) emerged naturally in different international conferences where such issues have been formally and informally discussed. This was the case, in the scientific field, at the *Symposium on Complex Bio Dynamics & Networks* (Tsuruoka City, 12–13 November, 2013) and in the philosophical field, among others, the SPSP conference (Toronto, Canada, 26–9 June 2013), the SPSP 2013 pre-conference: 'Science, Policy, Values: Exploring the Nexus', the ISHPSSB meeting (Montpellier, France, 7–12 July 2013), the EPSA conference (Helsinki, Finland, 28–31 August 2013), the Workshop on 'Synthetic Modeling of Life and Cognition: Open Questions' (Bergamo, Italy 12–14 September 2013), just to mention the events of the last year.

Some of the authors of this book met each other at the above conferences and enthusiastically adhered to an editorial initiative that hopefully records and makes explicit both their shared queries and their different perspectives.

ACKNOWLEDGEMENTS

We really thank Alfred Nordmann for having started a dialogue about the possibility to publish this work when the project was still at the very beginning.

I wish to thank the academic staff of the University Campus Bio-Medico and of the 'Institute of Philosophy of Scientific and Technological Practice'. I'm also very grateful to Sandra D. Mitchell, Kenneth Schaffner and Jim Woodward for the inspiring conversations we had during my fellowship at the Center for Philosophy of Science, Pittsburgh University, on these issues.

I also thank FCA of Verona for its support to the Bio-Techno-Practice project, and the project 'Models and Inferences in Science: Logical, Epistemological, and Cognitive Aspects', founded by the Italian Government (PRIN 20122T3PTZ) for contributing to the realization of this volume.

LIST OF FIGURES

LIST OF CONTRIBUTORS

Dino Accoto is assistant professor of Bioengineering at Università Campus Bio-Medico di Roma (Italy). He graduated in Mechanical Engineering from University of Pisa in 1998 and received his PhD in Biomedical Robotics from Scuola Superiore Sant'Anna in Pisa in 2002. In 2001 he was visiting researcher at Stanford University (CA, USA). From 2003 to 2007 he was Assistant Professor of Biomedical Engineering at Scuola Superiore Sant'Anna (Pisa, Italy). Since 2004 he is with the Biomedical Robotics and Biomicrosystems Laboratory, Università Campus Bio-Medico di Roma. His research activities focus on the development of biomedical robots and biomechatronic systems for assistance, rehabilitation and surgery, using a multi-scale and multi-domain approach. He co-authored about 100 peer-reviewed papers, and he is co-inventor of eleven patents. He is member of the ASME, of IEEE, of the Engineering in Medicine and Biology Society, and of the Robotics & Automation Society. He co-organized several international workshops and conferences special sessions on wearable robotics, tactile sensors and novel actuators. He is member of the editorial board of the Springer Springer Series on Biosystems and Biorobotics, and he served as special issue guest editor for *Computer Methods and Programs in Biomedicine*, *Robotics and Automation Magazine* and *Advances in Mechanical Engineering*.

Marta Bertolaso is Assistant Professor of Philosophy of Science at the University Campus Bio-Medico of Rome. She got a degree in biological sciences and she continued her academic studies with a PhD thesis in philosophy of life sciences. In 2012 she was Visiting Fellow at the Centre for Philosophy of Science of the University of Pittsburgh. In 2013–14, she was Visiting Fellow at the Centre of Biomedical Humanities of the Department of Experimental Oncology of IEO, University of Milan. She is particularly interested in the philosophical aspects of scientific practice and of the organizational dynamics of the biological systems. Causal complexity and functional heterogeneity in the biological systems, cancer research, the biology of cancer, and Systems Biology have been her main fields of inquiry in the last years. She has also published studies on ethical issues like the HET (Human Enhancement Technologies) and the concept of Quality of Life. She has published various papers in international journals and in different languages. Among her publications there are two monographs: *Il cancro come questione* (FrancoAngeli, 2012); *How Science Works: Choosing Levels of Explanation in Biological Sciences* (Rome: Aracne, 2013).

Federico Boem is a PhD student in 'Foundations of the Life Sciences and Their Ethical Consequences' at the European Institute of Oncology and University of Milan. Federico holds an Italian Laurea in Logic and Philosophy of Science (obtained at the University of Florence, Italy) and a Master in History and Philosophy of Biology (obtained at the University of Exeter, UK). He is specialized in philosophy of science, especially molecular biology and bioinformatics. Federico's research concerns the role of semantic tools (as Gene Ontology) in contemporary biomedical research in the context of data-driven versus hypotheses-driven science.

Giovanni Boniolo is Full Professor of Philosophy of Science and Medical Humanities at the Department of Health Sciences (University of Milan). He is the Director of the Biomedical Humanities Unit at the Department of Experimental Oncology of the IEO, and the Director of the PhD program in 'Foundations of the Life Sciences and their Ethical Consequences' www.ieo.it/it/RESEARCH/People/Researchers/Boniolo-Giovanni/

Marco Buzzoni is Full Professor of Philosophy of Science at the University of Macerata since 2001. He obtained a degree in philosophy at the University of Genoa (Italy) with a dissertation on 'Knowledge and Reality in K. R. Popper' (1979). He then obtained a postgraduate diploma in philosophy at the University of Genoa with a dissertation on 'Theory of Meaning, Ontology and Hermeneutics of Scientific Knowledge. Essay on T. S. Kuhn' (1984). From 1992 to 2001 he was Associate Professor of Philosophy of Science at the Institute of Philosophy at the Universities of Palermo and Macerata, and received an Alexander von Humboldt-Fellowship at the Universities of Würzburg, Marburg and Duisburg-Essen. He is an ordinary member of the Académie Internationale de Philosophie des Sciences and co-editor of the journal *Epistemologia: An Italian Journal for the Philosophy of Science*. His fields of research include Popper's and Kuhn's philosophy of science, science and technology, epistemology and methodology of human and social sciences, thought experiment in the natural sciences. His main publications are *Knowledge and Reality in K. R. Popper* (Milan, 1982); *Theory of Meaning, Ontology and Hermeneutics of Scientific Knowledge. Essay on Th. S. Kuhn* (Milan, 1986); *Operationalism and Hermeneutics: Essay on the Epistemological and Methodological Status of Psychoanalysis* (Milan, 1989); *Science and Technique: Theory and Experience in Natural Sciences* (Rome, 1995); *Experiment and Thought Experiment* (Milan, 2004); *Thought Experiment in the Natural Sciences* (Würzburg, 2008).

Nicola Di Stefano obtained a degree in philosophy at the University of Milan and in double-bass at the Conservatorio 'G. Verdi' of Milan. A PhD in bioethics followed at the University Campus Bio-Medico in Rome. He has epistemological and aesthetical interests. Currently he is focusing on consonance and dissonance perception in infants. He is a member of the 'Society for Music Perception and Cognition'. Recent publications include 'Nuages gris di Liszt: proposta d'analisi', *Rivista di analisi e teoria musicale*, 1 (2012), pp. 49–64, and (with G. Ghilardi), 'Embodied Intelligence: Epistemological Remarks on an Emerging Paradigm in the Artificial Intelligence Debate, *Epistemologia*, 1 (2013), pp.100–11.

Antonio Diéguez is Professor of Logic and Philosophy of Science at the University of Málaga, Spain. He was acting Dean and Vice-dean (1997–2002) of the Faculty of Philosophy and Humanities at the University of Málaga. He has been a visiting scholar at the University of Helsinki and at Harvard University. Currently, he is the first elected president of the Sociedad Iberoamericana de Filosofía de la Biología (AIFIBI). His first main research interest was the contemporary debate about scientific realism. On this issue he published a book – *Realismo científico* (*Scientific Realism*) (1998) – and more than fifteen papers. In these works he defended a moderate scientific realism. He has also published several papers and two books as a co-author on the difficulties to get a control of our modern technology. Afterwards, he extended his research interests to the field of Philosophy of Biology, working especially on the evolutionary explanationof the mind and its philosophical implications (evolutionary epistemology). He has published a book on this topic: *La evolución del conocimiento: De la mente animal a la mente humana* (*The Evolution of Knowledge: From Animal Mind to Human Mind*) (2011). He is also author of a handbook of philosophyof biology titled *La vida bajo escrutinio. Una introducción a la filosofía de la biología* (*Life Under Scrutiny. An Introduction to Philosophy of Biology*) (2012). He has published two other books: *La teoría de las ciencias morales en John Stuart Mill* (*John Stuart Mill's Theory of Moral Sciences*) (1988), and *Filosofía de la ciencia* (*Philosophy of Science*) (2005).

Giampaolo Ghilardi is a research fellow in the Institute of Philosophy of Scientific and Technological Practice at the University Campus Bio-Medico, in Rome. He graduated in Philosophy in Milan (Università Cattolica del Sacro Cuore), and Bioethics in Rome (Ph.D at University Campus Bio-Medico). He developed the notion of roboethics, and is currently studying the logic of robot design from an epistemological point of view.

Alessandro Giuliani is Senior Scientist at Istituto Superiore di Sanità (Italian NIH). For thirty years he has worked on the development of 'soft' quantitative modeling of biology and chemistry by means of multidimensional statistical analysis and network-based approaches. This allowed him to be the co-author of more than 200 papers in peer-review journals spanning from botany to biochemistry, behaviour, physiology, genetics, organic chemistry and toxicology. He was one of the inventors (together with Joseph Zbilut and Charles Webber) of Recurrence Quantification Analysis (RQA), a non-linear time series technique widely used in many fields of science. His main theoretical interest is in the development of a simple and versatile quantitative language for science focusing on the pattern of relations (law of form).

Wenceslao J. Gonzalez is Professor of Logic and Philosophy of Science (University of A Coruña). He is a Full Member of the *International Academy for Philosophy of Sciences* (AIPS). He has been a *Visiting Fellow* at the Center for Philosophy of Science (University of Pittsburgh) and a Team Leader of the European Science Foundation programme entitled 'The Philosophy of Science in a European Perspective'. He has been named a Distinguished Researcher by the Main National University of San Marcos in Lima (Peru).

Gonzalez has been a visiting researcher at the Universities of St Andrews, Münster and London (LSE). He has given lectures at the Universities of Pittsburgh, Stanford, Quebec and Helsinki. The conferences in which he has participated include those organized by the Universities of Uppsala, New South Wales, Bologna, Canterbury (NZ) and Beijing. He received the Research Award in Humanities given by the Autonomous Community of Galicia (Spain). He was President of the Committee of Doctoral Programmes at the University of A Coruña. He is the editor of the *Gallaecia Series: Studies in Contemporary Philosophy and Methodology of Science*, and he is co-editor of the *European Studies in Philosophy of Science*.

His publications include the edition of 36 volumes on philosophy and methodology of science, among them *Evolutionism: Present Approaches* (2008), *Scientific Realism and Democratic Society: The Philosophy of Philip Kitcher* (2011), *Conceptual Revolutions: From Cognitive Science to Medicine* (2011), *Freedom and Determinism: Social Sciences and Natural Sciences* (2012), and *Creativity, Innovation, and Complexity in Science* (2013).

Eugenio Guglielmelli (1991 MSc Electronics Engineering, 1995 PhD in Biomedical Robotics, both from the University of Pisa, Italy). He is currently Full Professor of Bioengineering at Campus Bio-Medico University (Rome, Italy) where he serves as the Head of the Laboratory of Biomedical Robotics and Biomicrosystems that he founded in 2004. From 2011 to 2013 he served as Director of Studies of the School of Engineering and he is currently serving as Pro-Rector of the Research at the same University. From 1991 to 2004, he worked at the ARTS Lab of the Scuola Superiore Sant'Anna (Pisa, Italy) in the group of Prof. Paolo Dario. His main current research interests are in the fields of human-centred robotics, biomechatronic design and biomorphic control of robotic systems, and in their application to robot-mediated motor therapy, assistive robotics, neuroengineering and neurorobotics. He is author/co-author of more than 170 papers on peer-reviewed international journals, conference proceedings and books. He was project coordinator of the FP7\FET-EVRYON project (www.evryon.eu, 2009–2012), and, since 1993, he participated in 20+ EU projects such as co-PI,

WP and Task Leader and Project Manager, etc. He is co-inventor of four patents and co-founder of four research spin-off companies. He currently serves as Associate Editor of the IEEE Transactions on Robotics, and as Editor-in-Chief of the Springer Series on Biosystems and Biorobotics. He is an IEEE Senior Member and he currently serves as Associate Vice-President for Membership Activities of the IEEE Robotics & Automation Society (RAS). In 2012, he also served as General Chair of the IEEE RAS\EMBS International Conference on Biomedical Robotics and Biomechatronics (BIOROB) and Program Chair of the IEEE\RSJ International Conference on Intelligent Robots and Systems (IROS). He served as independent expert reviewer and evaluator for EU FP6 and FP7.

Sui Huang, MD, PhD, obtained his degrees from the University of Zurich for work on interferon and did postdoctoral research at the Children's Hospital, Boston, on shape-dependent cell cycle control. He was faculty at Harvard Medical School and University of Calgary before joining the Institute for Systems Biology (Seattle). After more than a decade of research on endothelial cell fate control and tumour angiogenesis, Dr Huang has spearheaded the study of gene regulatory network dynamics and how it governs the generation of multiple cell types in metazoan. In his quest for a formal 'theory' of multi-cellularity, from which the natural inevitability of cancer may be derived, he has demonstrated that cell types are attractor states of gene regulatory networks and proposed a role for non-genetic ('mutation-independent') mechanisms in tumour progression. Dr Huang's lab currently studies cell fate decisions in terms of first principles of complex dynamical systems, embedded in cell population dynamics– a much neglected level of description between genotype and phenotype. One goal is to understand how such dynamics gives rise to the robust time-asymmetry of development and disease, notably, in cancer progression.

Cecilia Laschi is Associate Professor of Biorobotics at the Scuola Superiore Sant'Anna in Pisa, Italy, at the BioRobotics Institute, where she serves as Vice-Director. She serves as Rector's delegate to Research and PhD. She graduated in Computer Science at the University of Pisa in 1993 and received the PhD in Robotics from the University of Genoa in 1998. In 2001–2 she was JSPS visiting researcher at Waseda University in Tokyo.

Her research interests are in the field of biorobotics and she is currently working on humanoid robotics, soft robotics and neurodevelopmental engineering. She has been and currently is involved in many National and EU-funded projects, she was the coordinator of the ICT-FET OCTOPUS Integrating Project, leading to one of the first soft robot, and she coordinates the European Coordination Action on Soft Robotics RoboSoft. She has authored/co-authored more than 50 papers on ISI journals (over 200 in total), she is in the Editorial Board of *Bioinspiration & Biomimetics, Frontiers in Bionics and Biomimetics, Applied*

Bionics and Biomechanics, Advanced Robotics, and she has been Guest Co-Editor of Special Issues of *Bioinspiration & Biomimetics, Autonomous Robots, IEEE Transactions on Robotics, Applied Bionics and Biomechanics, Advanced Robotics*.

She is member of the IEEE, of the Engineering in Medicine and Biology Society, and of the Robotics & Automation Society, where she served as elected AdCom member and currently is Co-Chair of the TC on Soft Robotics.

Miles MacLeod is a current post-doctoral fellow of TINT Centre of Excellence in the Philosophy of the Social Science, Helsinki, Finland. He obtained his doctoral degree in history and philosophy of science from the University of Vienna (2010). Miles works principally on the subject of interdisciplinary model-building practices in science and in particular modern quantitative and engineering-driven biological fields such as systems biology. His analyses of scientific practices in these new fields draws from ethnographic studies of scientific research and scientific cognition performed in collaboration with Professor Nancy Nersessian (Georgia Institute of Technology). Results have been published in philosophy of science and cognitive science journals.

Alfredo Marcos originally studied Philosophy at the University of Barcelona. He completed his PhD on *'The Role of Information in Biology'*. He stayed at Cambridge and Rome. He has published a dozen books on Philosophy of Science, Environmental Ethics and Aristotelian Studies, and almost a hundred chapters and papers in journals of prestige, such as *Studies in History and Philosophy of Science* (Elsevier), *Information System Frontiers* (Kluwer) and *Science and Education* (Springer). He has taught courses and conferences at several universities in Spain, Italy, France, Poland, Mexico, Argentina and Colombia. Currently he teaches philosophy of science at the University of Valladolid (Spain), where he was Head of the Department of Philosophy, and he is currently Full Professor and Head of the PhD programme. Some recent publications include, as author: A. Marcos, *Postmodern Aristotle* (Newcastle: Cambridge Scholars Publishing, 2012); A. Marcos, *Ciencia y acción* (Mexico: FCE, 2010) (translated into Italian and Polish; second Spanish edn, 2013); as co-editor: S. Castro and A. Marcos, *The Paths of Creation: Creativity in Science and Art* (Berna: Peter Lang, 2011); and individual chapters: A. Marcos and R. Arp, 'Information in the Biological Sciences', in K. Kampourakis (ed.), *The Philosophy of Biology: A Companion for Educators* (Dordrecht: Springer, 2013), pp. 511–48; A. Marcos: 'Bioinformation as a Triadic Relation', in G. Terzis and R. Arp (eds), *Information and Living Systems: Philosophical and Scientific Perspectives* (Cambridge, MA: MIT Press, 2011), pp. 55–90.

Zsuzsa Pavelka is a PhD student in 'Foundations of the Life Sciences and Their Ethical Consequences' at the European Institute of Oncology and the University of Milan. She holds a Diploma in Biology from the University of Göttingen (Germany). Zsuzsa's project examines the use of genetic variants in clinical diagnosis, with focus on the establishment of pathological variants thorugh whole exome seqencing and in common diseases.

Kumar Selvarajoo joined the faculty at the Institute for Advanced Biosciences, Keio University, in April 2006. Currently, he is an Associate Professor and leading systems biology projects related to dynamic immune and cancer response. Prior to the relocation to Japan, Kumar was a Project Leader at the Bioinformatics Institute (A*STAR), Singapore. He obtained a Doctor of Philosophy (2004) for Computational Systems Biology at the Nanyang Technological University, Singapore, and a Master of Engineering degree (Aeronautics) at the Imperial College of Science, Technology and Medicine in London (1997). Currently, Kumar serves as an Editorial Board Member to *Nature*'s Scientific Reports, PLOS ONE and Frontiers in Systems Biology journals. He is also the editor-in-chief for the *Advances in Systems Biology* journal.

INTRODUCTION – PHILOSOPHY
WITHIN SCIENCE

The biological world has been always considered the locus of complexity. The fine structural and functional multi-layered organization of biological entities are immediately perceived as reaching a far higher level of integration than human artefacts. The huge technological advances of the last decades, however, have produced artefacts whose intrinsic complexity is, for many aspects, not totally manageable by the designers (in some cases acquiring a sort of independent existence, like world-wide-web), thus asking for an approach similar to the one adopted by biological sciences. It is now generally recognized that bio-related ideas play a key role in characterizing society, informatics and our daily life as well. In the last decades questions about the 'biological' have also been re-emerging in philosophy. The focus of the discussion has shifted from the ancient dichotomy of mechanistic versus holistic perspectives to the modern dichotomy of reductionist versus systemic views, when dealing with organisms and their dynamic behaviour. From the epistemological side we observe the progressive convergence of explanatory concepts in different fields and disciplines – from physics to physiology or ecology; from medicine to robotics. Such convergences affect functional accounts and the identification of explanatory systems. Moreover, common properties are acknowledged for different systems in terms of complexity, coherence, resilience, robustness, etc.

Familiar examples include the discussion and studies of how DNA expression is regulated, how epigenetic mechanisms influence developmental processes, how a system and its environment interact, and how a robot can display functional plasticity. However, the contribution that such pervasiveness of biological concepts makes to our understanding of the natural processes is not yet well understood.

Only recently have scientists started to look in-depth into the theoretical meaning and consequence of the use of biological-like explanations for artificial systems. Yet instead of bringing us back to the difference or reducibility among biological sciences and physics or chemistry, the increasing interconnection between technology and biology seems to prevent us from doing so. Such interconnection opens up new questions regarding causal accounts and the conceptualization of explanatory dynamics, as the first chapters of this volume show. The most relevant and important questions do not concern the relatedness of such fields, but the peculiarity of the scientific question that is addressed both in an experimental and technological field through biological concepts. For example, because organisms' parts are usually functionally defined, it would seem natural to associate functional explanation as peculiar to biological sciences, yet it is commonly used in robotics as well. Consistently, it would seem that teleological arguments might be an issue only for biological explanations while their role in mechanistic accounts still need further clarification and discussion.

We are witnessing now a continual interplay between different fields, which once had relatively sharp and clear-cut boundaries. These limits are rapidly fading. Philosophers and scientists have only gradually incorporated biological-related concepts in their methodology. This volume thus seeks to explore what we believe is a neglect of many discussions arising from such observations, particularly how the process of scientific understanding is shaped by our way of conceptualizing the world and how the interplay of 'biological' (we define with the acronym Bio), with technological (Techno) and philosophical-theoretical (Logos) ways of thinking are merging and contaminating their respective traditions.

Although there is apparently no single message or theme emerging from these pages, there is a shared commitment to articulating and addressing scientific phenomena. All the authors, in fact, agree that modern merging of the *bio* and *téchne* in engineering science is facilitated by new concepts and practices that can not be reduced to either *bio* or *téchne*, and can not be analysed using traditional philosophy of science dichotomies that roughly split artificial world and natural world. This is true whether talking about embodiment or stratification, for example. Understanding how this science of new concepts and practices works requires analysis and its own rationalization (Logos). Part I concerns the articulation of these boundary sciences, their concepts (perspectives) and how they work outside established science. Part II is about the epistemic analysis of these sciences and how they generate knowledge/understanding (and new technology as a result) in novel ways. And Part III is about their philosophical rationalization.

A strong unitary idea connects every different contribution, which, therefore, deals with the mutal and dynamic relationship between these three terms Bio-Techno-Logos. On the one side, 'Bio' can be seen as biological world, physical world or, in a general way, world of phenomena. On the other side, 'Logos' is the scientific understanding or representation of this world. In between *logos* and *bios*, *téchne* is how we conceive (*Logos*) natural world (*Bios*) and their mutual relationship. *Téchne* can be *hardware*, like robots or prosthesis, *software*, like computer simulation programmes and models, or purely *theoretical*, like scientific paradigms (complexity, systems biology, teleology, embodied intelligence).

Thus, Bio deals with the *what* question, Techno with the *how* question and Logos with the *why* question.

Part of the contribution of this volume is thus *methodological*, in the sense that we look forward to helping ferment an open discussion inside science using this tripartite perspective, which drives the discussion on every topic and which remains the very *leitmotiv* of the project. For this reason each author has been requested to pay particular attention to the way he expresses his concepts to keep the focus always on the aim of the volume rather than on the specific question each author discusses.

The chapters in this volume come from an international community of scholars who seek to present their own work from the perspective of the bio-techno-logos relationship. The term of 'logos' (instead of 'philosophy') has been preferred to highlight that, when considering such relationships, we are looking at different constitutive dimensions of the human understanding of the natural world. The emphasis is not on the different disciplines involved, but on how different tools (conceptual, technical, explanatory) are simultaneously involved in the process of scientific understanding that become, in our view, paradigmatic when the question is the dynamics of what we call living systems. Clearly, what all authors have in common is a strong interest in understanding scientific and philosophical trends of reflection and inquiry and to contribute through their own work to the development of new explanatory paradigms. All of them are, moreover, aware that human factors do affect scientific research although discrepancies and divergences on their relevance and nature appear as well. I am satisfied with such pluralism though integrated by an honest and opened willingness to listen to each other. The scientific biography and production of the authors might be taken into account as well in order to get a more comprehensive view of the issues at stake, thus moving further towards a wider discussion of the proposed topic: the bio-techno-logos relationship in and for scientific practice.

Authors' respective disciplines, which reveal the deeply interdisciplinary nature of these issues, include: biology, bioinformatics, chemistry, computer science, mathematics, systems biology, medicine, philosophy, physics and ethics. The contributions of the authors are intended to be useful not only to fellow researchers, but also to advanced undergraduate and graduate students in both science and philosophy. This volume therefore crosscuts different subjects contributing to a new field of contemporary debate about the relationship among scientific objects of inquiry, i.e., the relevance of technology as a way to better understand the world, and science in terms of *human practice* (see the final chapter).

Authors have been asked to put aside any disputes with the 'old way to do science' or even with the 'new way to do science' (i.e., reductionist vs anti-reductionist, molecular vs systemic, etc.). This is primarily an invitation to address the problems and issues of scientific research without necessarily applying to them the conceptual schemas of old vs. new contrapositions. More positively, authors have been asked to produce analyses and overviews in three areas of the epistemological and methodological reflection on biology. The three areas correspond to the three parts into which the book is divided: I. Biological dynamics; II. Reflecting on scientific understanding and understanding by building; III. Towards a development of a philosophy of scientific practice.

Part I: Biological Dynamics

Part I, therefore, mainly focuses on complex bio-dynamics and networks. It showcases some of the latest experimental and theoretical innovations addressing the complexity of responses in cell biology and the challenges and perspectives posed by bio-inspired robotics.

We are entering a new era in which biological sciences are adopting interdisciplinary strengths. Until the 1990s, steady-state, population averaged, single molecule in vitro approaches, in fact, have been dominating biological sciences. However, the field has progressed to track thousands of molecules dynamically within single cells using the next generation sequencing, high-throughput proteomics and metabolomics technologies. As more and more data are acquired, how do we understand the overwhelming information? In Chapter 1, 'Microscopic and Macroscopic Insights of Dynamic Cell Behaviour', Kumar Selvarajoo focuses on dynamical 'patterns' that can be observed from systems-level cellular information. By using distinct theoretical concepts, like landscapes and attractors, it is feasible to interpret complex behaviours of living systems at microscopic and macroscopic scales.

What kind of reflection arises when looking at something like proteins that are, in their own right, in the middle between chemical and biological worlds is extended by Alessandro Giuliani in Chapter 2, '"News from the Twilight Zone": Protein Molecules between the Crystal and the Watch'. He develops his own view regarding the bio-techno-logos relationship at that level. Proteins are the actual molecular 'players' of biological systems: the catalysis of metabolic reactions, the building of specialized structures, the specificity of immune response, and use is even made of the genetic information stored in DNA relying on the peculiar properties of proteins. These macromolecules live in a grey zone between chemistry and biology, allowing a unique perspective to the bio-techno-logos mutual relations. Proteins are in the same time molecules (and thus ruled by physical constraints) and very efficient microscopic machines (and thus driven by optimal task completion). This dual status makes protein features to keep track of both structural and functional optimization happening at different scales and in many occasions and following different goals. A crystal derives its structural features by the optimization of physical laws, on the contrary, the structure of a watch is almost totally determined by its function. Such features are related to the Logos and Techno concepts. Proteins are, in fact, just in the twilight zone between these two extreme paradigms, allowing us to derive precious hints about the mutual relations between the two by the intermediate effect of the third (Bio) term. Proteins are a perfect example of how the bio-techno-logos relationship can be strong: proteins *are* biological entities (Bios), they do act somehow in living processes (Téchne) and they deal with the deep reasons (Logos) of life development.

In Chapter 3, 'Limits to Deterministic-Linear Causality in Biomedicine: Effects of Stochasticity and Non-Linearity in Molecular Networks', Sui Huang discusses how in modern Western sciences we expect 'arrows of explanation' to point 'downwards' in that there is causation at a more 'fundamental' level, and how biomedical sciences is no exception in this environment. We seek to understand diseases, or more generally, phenotypes (the macroscopic observable in biology), in terms of molecular pathways. But in addition to the cultural-epistemic reasons, this self-imposed limitation on the proximate molecular cause of a phenotype has a practical justification: molecules are the targets of therapeutic intervention, the handles by which drugs act to steer phenotype into a desired direction. However, in establishing the molecular causation one rarely takes into account a set of elementary properties immanent to complex systems to which multicellular organisms belong. These characteristics entail a departure from our intuition of causation, interpreting its bottom-up arrow's direction in an non-obvious way, and include: (i) stochasticity of processes and events, discussed in the previous chapter, (ii) non-linear dynamics of molecular and cellular networks. Prof Huang takes into account the implications of networks with their stochastic non-linear dynamics on the habitual notion of causality in biology, which assumes a deterministic and linear relationship between cause and effect. He introduces with sufficient detail but in a manner that can be comprehended with qualitative reasoning and with minimal mathematical formalism several central concepts, such as multi-stability and bifurcations. These are central to understanding departures from deterministic, linear causality. He discusses how they explain counterintuitive phenomena that defy comprehension in terms of the familiar deterministic linear cause–effect relationship.

Finally, in Chapter 4, 'Embodied Intelligence in the Biomechatronic Design of Robots', Dino Accoto, Eugenio Guglielmelli and Cecilia Laschi introduce the reader into the field of biomechatronics. It is the elective design approach in the development of complex machines having a strong bond with the biological world. Biomechatronics revolves around the concept of convergent design of mechanical and electronic subsystems, with a constant attention to the biological component. As is customary in engineering disciplines, the design phase, especially if oriented to achieve the optimal solution for a particular engineering problem, is rooted in system models. In the case of biomechatronic systems, such models are inherently *multidomain*, as they involve different branches of knowledge, with particular emphasis on mechanics, electronics and life sciences. The biological component and the environment in which the machine operates are traditionally considered as factors outside the domain of the project in which the designer has freedom of action. But this view underwent a major change in recent years. In fact, it was shown that the dynamic interaction, established among the machine and these factors, can lead to – in some meanings of the word – 'emerging' behaviours that may be useful for achieving the design goal. This evidence represents the foundation of the so-called Embodied Intelligence paradigm. Designing a machine so that it is able to elicit emerging behaviours, obtained not through *calculation* but through *morphological computation*, requires specific operational tools, the definition of which is the subject of several research lines. To illustrate the state of the art in this area, in the chapter they examine two representative case studies, focused on the development respectively of soft bio-inspired robots and wearable robots for the assistance of disabled people. In this context, biological world (Bios) is a model for inspiring robotics (Téchne) for which embodied intelligence is the adequate paradigm (Logos) for better understanding the system–environment relationship.

The discussion of the links between models, design methods and technology highlights how the performance optimization, based on the exploitation of the dynamical interaction among machine, biological component and the environment, poses challenges that can hardly be addressed by only resorting to the analytical and reductionist approaches developed over the last four centuries. Such effort clearly opens the discussion for papers in Part II (the chapter by Ghilardi) and Part III (the chapter by Dieguez and Gonzalez) that deals with models and explanatory tools in technoscience.

Part II: Reflecting on Scientific Understanding and Understanding by Building

The second part highlights the main challenges of the explanatory enterprise through technology when the scientific question is on dynamics of biological systems. These chapters introduce the idea that innovative aims and methods of model-building in systems biology – in first place understood as a discipline that deals with integrated systemic dynamics – do open new challenges for philosophy of science and suggest different perspectives through which complexities of such dynamics can be managed. Integrations of different levels, overall view, identification of the different features and components of new technological applications seem to be the key points chosen to reach a better understanding and assessment of biological research. Given the general inspiration of the volume, this approach promises to develop a more satisfactory epistemology of biological sciences.

Miles MacLeod, in Chapter 5, 'Managing Complexity: Model-Building in Systems Biology and its Challenges for Philosophy of Science', first gives an overview of some of the new scientific practices computational systems biology is creating through its particular merging of technology, quantitative methods and concepts into biological investigation. Drawing upon a five-year ethnographic study of two systems biology labs, he suggests that current philosophical analyses of the field do not capture the rich interdisciplinary practice and pragmatic engineering mindset that compose systems biology, limitations that partly result from relying on philosophical analysis that work well to characterize biology and biological practices but not necessarily this integrated bio-engineering context. The *mesoscopic* feature of scientific research emerges here in relation to systems biology. He also clearly points out that, within this field, such mesoscopic research – rather than top-down or bottom-up – is divided over whether the principal goal of systems biology is to build a mathematical theory of biological systems, which has consequences for laboratory research practices. Different meanings and uses of the mesoscopic term are also highlighted. Therefore, in general, systems biology cannot be analysed as a 'normal science' structured by problem-solving routines and standards but as one that depends on methodologically flexible and pragmatic responses to complexity. He suggests that mechanistic explanation may not always be the best framework for characterizing the epistemic goals of systems biologists who often pursue more pragmatic goals that facilitate mathematical abstraction and mathematical short cuts in place of having clear mechanistic pictures. These areas suggest directions philosophers need to go to understand how the engineering mindset and quantitative approaches reshape the possibilities of biological investigation. The Bio-Techno-Logos triad, here, is interpreted in a new and interesting perspective in which mathematics and computational simulation are the tools (Téchne) through which scientific community addresses questions raising from biological world (Bios).

Explanation, causes and scientific understanding will be also discussed, from an epistemological perspective, by Buzzoni in Part III and with some interesting observations to the propositional aspect of knowledge by Dieguez.

In Chapter 6, 'Stratification and Biomedicine: How Philosophy Stems from Medicine and Biotechnology', moving from a different perspective to that which Huang offered in Chapter 3, Federico Boem, Giovanni Boniolo and Zsuzsa Pavelka present recent advances in molecular biology and biotechnology and show how they can change our perceptions of disease, diagnosis and therapy. As usually happens in the history of science, this innovative turn in medicine has begun spurring new philosophical analyses. In this chapter the authors focus on classification, through the complex concept of stratification of diseases, therapies and patients and on what stratification implies from a philosophical point of view. They start by describing the "stratification", its novelties and its impact on methods of treating patients in clinical contexts. Subsequently they show the philosophical implications stratification has. The Logos, here, meets the biological dimension of reality (Bios) through the importance of the notion of Bio-Ontologies, as the authors show.

In Chapter 7, 'Epistemology of Robotics: An Outline', Giampaolo Ghilardi, in dialogue with chapter four, highlights the epistemological contents of the relationship between humans (Bio) and robots (Techno). New technology has made, in fact, conceivable new designing trends. The previous division between software and hardware has been softened, so it is possible to talk about smart materials and embodied intelligence, changing the classical model of thinking about Artificial Intelligence. The paradigm shift between a top-down way of designing machines has changed the conception of robots itself, which are now considered moral agents. Agency applied to robots is discussed, aiming to define to what extent it is possible to adopt this image when talking about robotics.

Part III: Towards a Development of a Philosophy of Scientific Practice

In Part III we outline two main epistemological issues at stake in scientific explanations that still require philosophical reflection, and offer answers to the question about the nature of the entanglement of human and technological factors in the process of scientific understanding. We also present a theoretical framework of how the Bio-Techno merging requires new accounts of scientific practice.

In Chapter 8, 'Prediction and Prescription in Biological Systems: The Role of Technology for Measurement and Transformation', Wenceslao J. Gonzalez addresses complexity as one of the features of biological systems. Complexity is twofold: (1) it can be structural, when the biological traits are associated with the configuration of a system of such a part of natural reality; and (2) it can be dynamic, when the characteristics of the system evolved over time, which leads to new aspects of the biological system.

Prediction occurs in biological sciences insofar as they are basic sciences, but prediction also has a role in the biological sciences as applied sciences. In this regard, prediction can foretell some traits of the biological systems from the point of view of its structural complexity, and it can also predict the characteristics of a system that evolves over time. Prescription, as distinct from prediction, also has a role in biological sciences as applied sciences by identifying the constraints that can lead the biological system to a specific configuration. It can also offer some patterns on how the biological system should evolve over time. Both epistemological and methodological components of prediction and prescription require the use of technology. The role of technology is clear regarding the tasks of measuring the biological reality available, both in structural terms and in dynamic terms. In addition, technology also has a role in the transformation of the biological reality according to some prescriptions (such as the protection of certain natural environments).

In Chapter 9, 'Teleology and Mechanism in Biology', Marco Buzzoni approaches the topic of teleology and mechanism in biology. The main exponents of the mechanistic approach dispense with the explicit reductionist strategy of nineteenth-century mechanism and the positivist unity of science, but their silence on teleology shows that final causes are considered useless and/or redundant in biological investigations. But the concept of mechanism is one-sided and incomplete without an implicit reference to intentional and final causes, as is obvious considering the concept of mechanism or machine in the perspective of the manipulative theory of causality. In the third section, there is a step in the direction of authors who have tried to reconcile teleology and mechanism. There is room to develop a more satisfactory account of biological practice not only from elements of the mechanical viewpoint, but also referring to intentional or final causes. The reflexive, typically human concept of *telos* (Logos) may be employed successfully to investigate living beings (Bios) scientifically in an intersubjectively testable and reproducible way, to discover mechanical-causal or experimental relations in living systems.

In Chapter 10, 'Scientific Understanding and the Explanatory Use of False Models', Antonio Diéguez explores the scientific understanding and the explanatory use of false models, noting that in model-based sciences, like biology, models play an outstanding explanatory role. In recent times, some authors have shown how the notion of understanding could shed light on the analysis of explanation based on models, deepening the notion of causality that has also been addressed by Buzzoni and MacLeod. This notion has attracted growing attention in philosophy of science. Three important questions have been central in the debate: (i) What is scientific understanding? (ii) is understanding factive, i.e. does understanding presuppose or imply truth? and (iii) can understanding be objective? Diéguez outlines and assesses the main answers to these questions, thowing his own support behind positive answers to both questions ii and iii. Understanding can be factual and objective. But to be so, scientific understanding and explanation (Logos) depend on the elaboration of models (Téchne) of reality (Bios).

Finally, in Chapter 11, entitled 'Bio-Techno-Logos and Scientific Practice', Marta Bertolaso, Nicola Di Stefano, Giampaolo Ghilardi and Alfredo Marcos consider the role of the personal agency in scientific practice articulating the reflection on scientific understanding and practice and taking into account the processes of conceptualization and of identification of the explanatory levels. In particular they consider the fact that we can explain nature according to what the world allows rather than what pure logic requires. In this chapter a different perspective on science and of the scientific work as human action is therefore considered. What, in fact, makes lab works scientific is not just the 'scientific method' understood in terms completely grounded on observation and rational inference. This shift opens a new point of view, which is based on epistemological virtues, by means of which we can appreciate the multidimensional character of scientific practice.

Final Remarks

The starting point of our reflection has been that in recent years important philosophical issues and questions have emerged from science itself. The questions I have mentioned in the Preface have been driving discussions and the choice of the topics. No one really has a clear idea of what effect bio – or techno – has on technological or biological fields. Until we do, we do not know what conceptually follows from sticking a 'bio' on the front of, for example, engineering. Understanding why the prefix 'bio' is becoming so common in different fields, from informatics to robotics, and what kind of methodological and philosophical issues are at stake when we focus on 'biological' dynamics, clearly asks for a deeper philosophical reflection.

This encourages a reflection on philosophy of science and scientific practice as well. The strategy we adopted has dealt with questions about the peculiar dynamic features of complex systems we identify with the adjective of 'biological'. At the intersection of these philosophical and scientific topics ontological and epistemological issues emerged that, at the same time, have been clarifying what is at stake in the intrinsic relationship among the bio-techno-logos dimensions of scientific knowledge and what kind of rationality scientific understanding implies.

The shared theoretical background that supports the coherence of the book is reflected in the three parts. We are convinced that philosophy of science and especially philosophy of biology is entering a new phase. Until now philosophers interested in biology, as well as biologists with an interest in philosophy or, more generally, with a speculative interest in their work, kept separate the areas of speculation and the 'bench' work. In other words, the interest was in the 'general consequences' of scientific progress in biology (e.g. evolution theory, neuroscience, ecology, genetics and their impact on society or on long time-honoured philosophical themes like free will or consciousness) or on the epistemological status of biology with respect to classical categories like determinism, probability, reductionism vs holism, and so forth.

Things are rapidly changing and, as often happens, the change is driven by the widespread sensation that we are in the middle of a deep knowledge crisis.

The number of new drugs (in terms of active principles) entering the market has been steadily dropping for about thirty years. The 'invasion' of new data from high-throughput methods asks for something very different from 'smart' algorithms that, while putting some order on data, are not sufficient to generate new theoretical perspectives. The increasing fragmentation of knowledge is producing incommensurable results that are practically impossible to gather in a common frame.

To overcome the sensation that we are missing some very basic points and that the mere accumulation of new data can only worsen the situation, we decided to stop and think, not to struggle to have our 'particular point of view' prevail. What emerges from the volume confirms the conviction reflected in the title of the volume: that living matter, technology and the abstract, formal thinking have some irreducible peculiarities we can roughly summarize in three modes that are progressively developed in the three parts: (1) the 'organic' mode (a global entity whose behaviour far exceeds the sum of its parts that – in turn – are difficult to isolate); (2) the 'integrated' mode (a tool that works by connecting and directing its elements to a well-defined single goal); and going beyond, (3) the 'symbolic' mode (a description that satisfies some abstract and self-consistent rules of composition). These modes follow different ways of 'functioning' and obey different laws so an element of the 'organic' world can hardly be constrained into a fully consistent set of rules and the production of technically effective tools can hardly be reduced to pure logic while activities like drug design cannot be faced by pure engineering. On the other hand, bio-inspired engineering works very well, logical analysis of biopolymers like DNA and protein molecules allowed us to discover many facts of nature and the circuits of our desktop are a material instantiation of millennia of logic speculations.

Authors have been challenging themselves trying to clarify the ingredients that allow these three separated worlds to interact fruitfully and the misconceptions make the connections among these worlds difficult. They tried to do this in dialogue with other authors and disciplines while indicating where they see a 'bright light' in the distance. This justifies the different perspectives present here, with a wide overview on the most relevant points in this reflection opening further discussions that can develop from this first collection of essays.

Finally, the use we make of the classical *bios, téchne, logos* concepts is therefore justified by the belief that part of the answer is already posed in the question, i.e. in some intrinsic relationship among these concepts already acknowledged in the Ancient Greek representation of nature. This ancient relational view of those concepts can be now grasped in different ways through technological applications in biological sciences, and seems to be better suited to modern scientific practices, avoiding the risk of getting stacked in useless dichotomies (typically, the part-whole issue, top-down and bottom-up causation, etc.). Therefore, there is not only the need to escape traditional philosophical distinctions or over-representations, such as representing nature as a machine, but also the interesting focusing on their relationships, as the ancient Greek philosophers attempted to do.

1 MICROSCOPIC AND MACROSCOPIC INSIGHTS ON DYNAMIC CELL BEHAVIOUR

Kumar Selvarajoo

High-Throughput Biology

Biological systems, for the goal of survival, not only depend on external food sources, but also display resilience to pathogenic attacks or diverse environmental changes. The external contact, or perturbations, are continuously processed intracellularly through different levels of molecular interactions. Due to the enormous challenges facing survival, a living cell contains several thousands of genes, proteins and metabolites that are highly interconnected, leading to complex networks that can invoke highly sophisticated and variable responses.

To comprehend the complex system-level properties of living systems, over the last few decades the development of high-throughput experimental technologies for genomics, proteomics and metabolomics have proceeded intensively. These methods generate a large quantity of biological data at different scales crucial for unravelling the detailed molecular composition and complexity of living organisms.

For instance, what types of genes and how many of them are induced by the immune cells in response to invading pathogens? In a study to investigate the dynamic response of innate immune cells (macrophages) exposed to pathogenic agents (lipopolysaccharide), cDNA microarrays revealed that almost 3,000 genes, belonging to diverse cellular processes, were expressed over a period of 24 hours.[1] These genes were not only all related to the immune process, but also consisted of diverse cellular processes such as cell differentiation and cell cycle with distinct temporal responses. Such valuable information suggests that high-throughput analysis of biological components are crucial for understanding complex cell behaviours. As more and more studies focused on high-throughput data generation, system-level properties of living organisms were increasingly revealed.

Scale-Free Organization

In the early part of last decade, one study investigated the protein–protein interaction (PPI) networks of *Saccharomyces cerevisiae*, by searching for a relationship between the number of protein species and protein interaction causality.[2] Notably, the investigations showed that the probability of a protein interacting with *k* other proteins followed a statistical power-law. This feature, which is commonly found in social behavior, finance and the internet, was subsequently also shown for other biological species.[3] That is, a few highly connected individuals (proteins) play a central role in mediating interactions among numerous, less connected individuals (proteins), a property of scale-free network. This postulation suggests that biological networks are not connected randomly, but centres around a small proportion of 'hub' elements.[4] Thus, the removal of a relatively few highly linked 'hub' molecules can lead to system failure. On the other hand, targeting the 'node' (less connected species) will not produce significant changes to the system.

The activation of tumour-necrosis factor (TNF)-receptor-associated factor 6 (TRAF6) occurs in several distinct signalling cascades. Targeted disruption of TRAF6 in mice led to systemic failure and premature death,[5] suggesting that it is a 'hub' element. Similarly, 'hub' molecules such as *cycA*, *cdk2* and *p53* were also identified in p53 network, a well-known tumour suppressing pathway that control apoptosis, cell cycle, etc.[6] Therefore, catastrophic failure can occur due to the loss of function of 'hub' molecules, though the interference of 'nodes' will not show a big difference.

Overall, it is now convincing that the scale-free design of biological networks produces the inherent robustness and fragility to distinct environmental threats. Thus, the high-throughput technologies have been crucial for revealing global properties and organizing principles. However, the comprehensive understanding of dynamic cellular behaviours and their control mechanisms still remains a big challenge. For example, how does a certain deterministic cell differentiation trajectory deviate into a cancerous or disease state over time? To put high-throughput data together so that dynamical properties can be better understood requires the integration of theoretical and computational approaches with experimental dataset.

Recently, there has been active development of systemic methodologies to interpret dynamical cell behaviour. This phenomenon has led to the creation of a new interdisciplinary field, called systems biology, inviting scientists across various fields to actively participate in joint research. Systems biology methods are based upon formalized theories, in most cases utilizing physico-chemical laws, which may combine with spatial-temporal high-throughput experimental data to provide better insights into the underlying molecular circuitry that controls complex behaviours of living systems.

Modular Systems Biology

To consider a living system in its entirety, though desirable, is an overwhelming task. Therefore, to reduce the complexity, it would be appropriate to modularize cellular systems. Functional modules are compartments which can be considered as a workable system in isolation and has been widely adopted in engineering. In biology, this concept has been inherently observed for many years through in vitro experimental conceptualization. This has occurred, for example, in the modularization of the gene regulation system for the determination of transcriptional machinery, signal transduction cascades for the understanding of extracellular signal propagation into the nucleus, and metabolic pathways for evaluating the distribution of fluxes to a given concentration perturbation (see Figure 1.1). The ability to create functional modules through synthetic biology has become popular in recent years.[7]

The use of simple modularized dynamic models of cell signalling have produced very useful insights in the regulation of key transcription factors and their gene expressions. In one study, to elucidate the mechanisms behind distinct gene expression programmes triggered by the same IKK–IκB–NF–κB signalling module for different stimuli in mice, a highly simplified model was developed.[8] A mass-action response equations model, with twenty-four molecular species and seventy reactions for the dynamic response induced by both tumour-necrosis factor (TNF) and lipopolysaccharide (LPS) stimulation, suggested that the mechanism for the rapid termination of IKK activity in TNF stimulation is due to negative feedback control by post induction of A20, while the prolonged activation of IKK observed in LPS stimulation is due to positive feedback of autocrine TNF signalling. Although several hundreds of proteins are activated in TNF and LPS stimulation, it is remarkable that a highly abstracted simple model was able to pinpoint an important regulatory mechanism in a perceived complex mammalian immune response.

In a similar way to the IKK–IκB–NF–κB modular approach, another study used a computational model based on modular response approach (MRA) to reveal distinct wiring of the conserved MAPK signalling to epidermal growth factor (EGF) and nerve growth factor (NGF) stimuli.[9] MRA is based on ordinary differential equations with steady-state assumptions of biochemical networks over time. A matrix of (local) response coefficients is quantified to assess the nature of inter-modular interactions.[10] The use of MRA interpreted the experimental profiles where EGF activates ERK transiently through negative feedback from ERK to RAF, and NGF sustains ERK activity via positive feedback using the same axis.

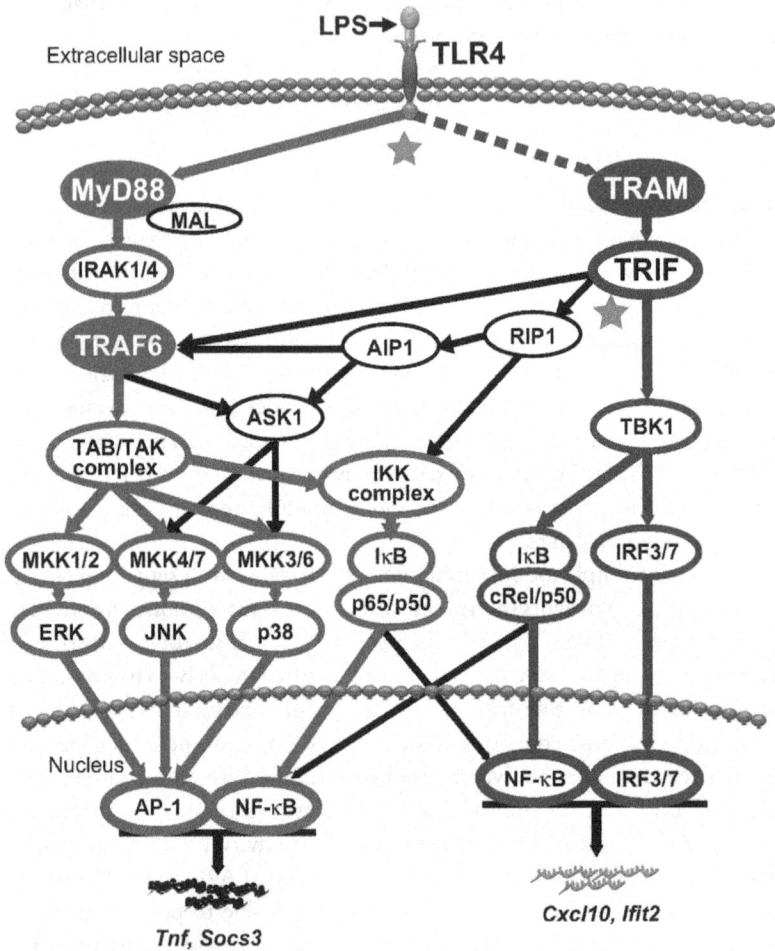

Figure 1.1: The functional modularization of biological networks: the Toll-like receptor 4 (Source: K. Selvarajoo, Y. Takada, J. Gohda, M. Helmy, S. Akira, M. Tomita, M. Tsuchiya, J. Inoue and K. Matsuo, 'Signaling Flux Redistribution at Toll-Like Receptor Pathway Junctions', *PLoS One*, 3 (2008) ep. 3430).

Thus, from the systems biology studies of IKK–IκB–NF–κB and MAPK signalling modules, it is clearly evident that despite the vast complexity of biological networks within living cells, deterministic immune responses can be predicted using modularized computational models. In other words, although a large number of molecules are packed within cells giving thoughts for a vast combinatorial possibility of unpredictable outcomes, biological responses for a given stimulus can occur through specific functional modules obeying simple and deterministic mass-action rules.

Using the modularization concept, to date, distinct mathematical and computational methodologies have been developed and studied for almost all known biological networks. Popular quantitative approaches include kinetic methodologies based on chemical equilibrium and the law of mass-action to analyse the dynamic response of biological networks,[11] multi-scale modelling that integrates the interactions between different levels of spatial and temporal organization for understanding system behaviour,[12] stochastic models that investigates the temporal probability of switching one state to another, or remain in the native state, especially for cell fate decisions.[13] Other approaches include stoichiometric methods, which deal with reaction stoichiometry to identify reaction mechanisms,[14] singular-value decomposition methods and statistical clustering for visualizing and identifying coordinated self-organizing behaviour from high-throughput data,[15] and Boolean and Bayesian network approaches for constructing biological network connectivity using discretization methods and probabilistic graphs.[16]

The consideration of functional modules in biology has led to the simplification of cellular response and has revealed novel network causality. This may not be surprising as several other fields such as physics, chemistry and engineering also display modular characteristics where without knowing most details of system architecture or parameter values, very useful knowledge about system dynamics are revealed. The law of nature has been shown to constrain the system so that even with limited information useful hidden properties can be revealed. For example, without knowing the details of wind dynamics, the behaviour of an aircraft in diverse atmospheric conditions can be predicted combining the laws of mass, momentum and energy conservation. Although such macroscopic laws will break down in certain conditions, such as in turbulent flow, or high speeds (greater than the speed of sound) when secondary effects such as kinetic heating complicate matter, nevertheless, the success of air and space travel is primarily due to the discovery of physical laws. Do such laws of nature also operate in biology?[17]

The Linear Response Approach

Interestingly, the dynamical profiles of several intracellular molecules activated through biological networks display linear response. Figure 1.2A–B show two examples taken from TLR3 and TLR4 signalling studies. In other words, the response profiles of complex biological networks can be delineated by the formation and depletion response waves.[18]

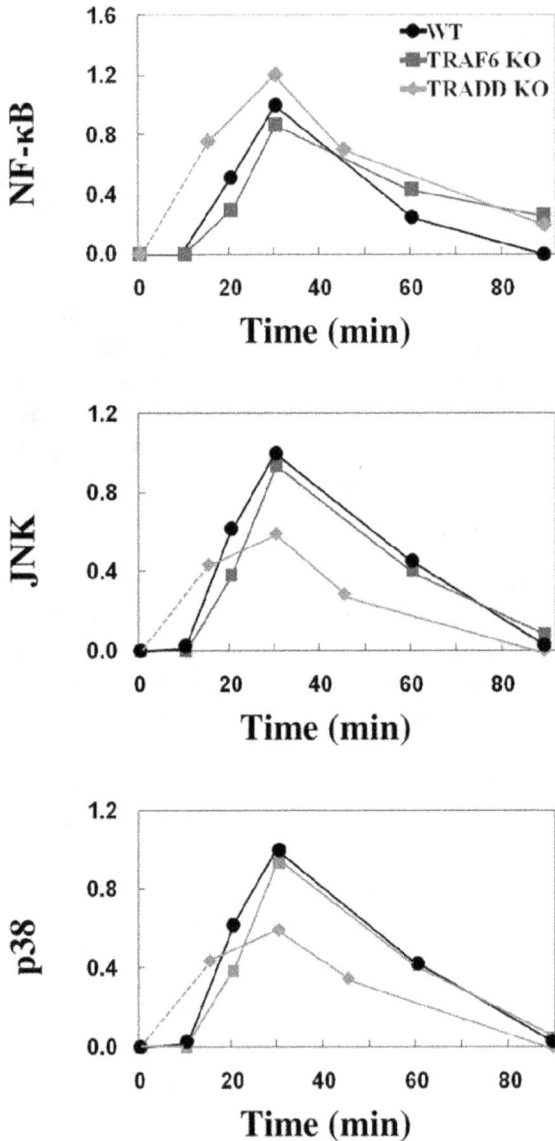

Figure 1.2A: The response of biological pathways to upstream perturbation shows deterministic downstream formation and depletion waves. TLR3 signalling dynamics show activation and deactivation following formation and depletion waves. Source: M. Helmy, J. Gohda, J. Inoue, M. Tomita, M. Tsuchiya and K. Selvarajoo, 'Predicting Novel Features of Toll-Like Receptor 3 Signaling in Macrophages', *PLoS One,* 4 (2009), ep. 4661.

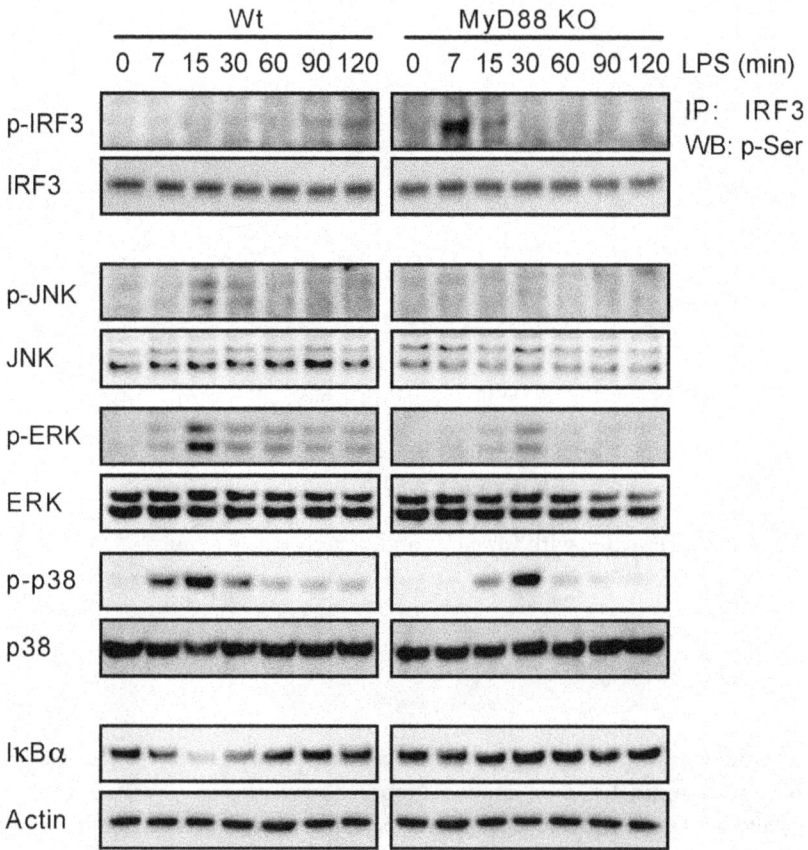

Figure 1.2B: The response of biological pathways to upstream perturbation shows deterministic downstream formation and depletion waves. TLR4 signalling dynamics show activation and deactivation following formation and depletion waves. Source: K. Selvarajoo, Y. Takada, J. Gohda, M. Helmy, S. Akira, M. Tomita, M. Tsuchiya, J. Inou and K. Matsuo, 'Signaling Flux Redistribution at Toll-Like Receptor Pathway Junctions', *PLoS One*, 3 (2008) ep. 3430.

To understand why biological network dynamics can follow linear response (or the superposition of formation and depletion response waves), let us consider pulse perturbation $(\alpha, 0)$ at $t = 0$, given to a simple two-species chain governed by first-order mass-action equations; $X = (X_1, X_2)$: $X_1 \xrightarrow{k_1} X_2 \xrightarrow{k_2}$. The perturbation wave $\delta X = (\delta X_1, \delta X_2)$, applied to the system with rate constants k_1 and k_2 for X_1 and X_2, can be represented by:

$$\frac{d\delta X}{dt} = \begin{pmatrix} -k_1 & 0 \\ k_1 & -k_2 \end{pmatrix} \delta X \tag{1}$$

With initial conditions, X_0, solving Eq. 1 yield the sum-of-exponentials:

$$\delta X_1 = \alpha e^{-k_1 t} \tag{2}$$

$$\delta X_2 = \frac{k_1}{k_2 - k_1}(e^{-k_1 t} - e^{-k_2 t}) \tag{3}$$

Factorizing Eq. 3 with respect to : $e^{-k_1 t}$, if $k_2 > k_1$ (or $e^{-k_2 t}$ if $k_1 > k_2$):

$$\delta X_2 = \frac{k_1}{k_2 - k_1}(1 - e^{-(k_2-k_1)t})e^{-k_1 t} \tag{4}$$

Eq. 4 can be rewritten with *formation* and *depletion wave terms*:

$$\delta X = \alpha \times \underbrace{(1 - e^{-P_1 t})} \times \underbrace{e^{-P_2 t}} \tag{5}$$

Perturbation   Formation wave   Depletion wave

where δ represents the amount of perturbation, p_1 and p_2 represent the measure of formation and depletion response propagation waves, respectively. In general, p_1 and p_2 are not equal to k_1 and k_2 respectively and are determined by fitting with experimental data. Thus, the origin for the formation and depletion terms observed for several biological network responses (Figure 1.2) is likely to have their origins from linear response equations.

The Origins of Linear Response

To investigate deeper into the origins of formation and depletion waves in the response of various complex biological networks outlined so far, let a stable network consisting of n species be perturbed from the reference steady-state. In general, the resultant changes in the concentration of species are governed by the kinetic evolution equation:

$$\frac{\partial X_i}{\partial t} = F_i(X_1, X_2, ..., X_n), i = 1, ..., n \tag{6}$$

where the corresponding vector form of Eq. 6 is $\frac{\partial X}{\partial t} = F(X)$. F is a vector of any non-linear function which can include diffusion and reaction of the species vector $X = (X_1, X, \dots X_n)$, representing the activated concentration levels of all reaction species. The response to perturbation can be written by $X = X_0 + \delta\mathbf{X}$, where X_0 is the reference steady-state vector and $\delta\mathbf{X}$ is the relative response from steady-states ($\delta X_{t=0} = \mathbf{0}$).

The generally non-linear kinetic evolution Eq. 6 can be approximated or linearized by using the Taylor series:

$$\frac{\partial \delta X}{\partial t} = \frac{\partial F(X)}{\partial X}\bigg|\delta X + \frac{\partial F^2(X)}{\partial X^2}\bigg|\delta X^2 + \dots \tag{7}$$

As the general volume of perturbing substance is usually very small (order of 1%) compared to the total volume of cells that are perturbed, now consider a small perturbation around the steady-state in Eq. 7, in which higher-order terms become negligible and results in the approximation of the first-order term. In vector form $\frac{d\delta X}{dt} \cong \frac{\partial F(X)}{\partial X}\big|_{X=X_s} \delta X$ (note the change from partial derivative to total derivative of time), where the zeroth order term $F(X_0) = 0$ at the steady-state X_0 and the *Jacobian* matrix, or linear stability matrix, is $J = \frac{\partial F(X)}{\partial X}\big|_{x=X_s}$. The elements of J, based on the initial activation topology, are chosen by fitting δX with corresponding experimental profiles. Hence, the amount of response (flux propagated) along a biological pathway can be approximated using *first order mass-action response*, i.e. $\frac{d\delta X}{dt} = J\delta X$. That is, the basic principle so far suggests that the response rate of species in a mass-conserved system at an initial steady-state can be approximated by first order mass-action response equation, given a small perturbation to one or more species.

Note that *Jacobian* matrix elements (or response coefficients) can include not only reaction information, but also spatial information such as diffusion and transport mechanisms. Thus, each species in the perturbation-response model can represent a molecule, a different modified state of a molecule (e.g. ubiquitinated state) or a molecular process such as diffusion, endocytosis, etc. That is, each species in the biological network does not necessarily represent a specific molecular species. Thus, unlike bottom-up kinetic models, which use fixed network topology, the perturbation-response approach considers the network as a sequence of events rather than just molecules. As molecular networks are largely not fully understood, this difference is crucial as it prevents rigidly fixing the network topology, and allows it to be modified according to experimental data so as to prevent over-fitting problems and to identify novel features of biological networks.[19]

Using the linear response approach, we have investigated the dynamic cell signalling of immune cells to proinflammatory stimulation and cancer cells to immunotherapy. To date, we have developed dynamic computational models based on the toll-like receptors (TLRs), tumour-necrosis factor (TNF) receptor, and TRAIL receptor stimulations. Our population average models have been successful in predicting novel properties of the signalling networks, and have been experimentally validated. For example, our TLR4 signaling model predicted (1) the presence of novel signalling intermediates along the TRAM-dependent pathways,[20](2) crosstalk mechanisms between the MyD88- and TRAM-dependent pathways,[21] and (3) the concept of *signalling flux redistribution* or *SFR*.[22] These results have been verified experimentally.[23] We have also used the concept of *SFR* to identify a crucial molecular target to enhance cellular apoptosis in cancer cells.[24] More recently, we have used the linear response model to predict computationally and validate experimentally a crucial target for reducing TNF-induced proinflammatory condition.[25]

Stochasticity and Variability from Single Cell Dynamics

The population-wide averaging approaches, discussed so far, have been instrumental, not only for cell signalling, but also in our basic understanding of myriad deterministic biological processes such as growth and metabolism. However, each cell within a population is not identical in its morphology or shape, and the intracellular molecular environment is highly inhomogenous. Furthermore, we now know that even genetically identical cells produce diverse phenotypes: a single stem or progenitor cell can produce distinct lineages, which can be tilted even by small external perturbations. Population-wide average techniques or linear response methods are, therefore, not suitable for understanding cellular variability and inhomogeneity. Thus, the understanding of how multimodal cellular decisions can be undertaken by living systems, e.g. in the multifaceted outcomes of stem cell differentiation, requires non-averaging and non-linear approaches considering variability.

   Since the beginning of the millennium, experimental techniques have progressed towards addressing the issue of heterogeneity and their implications from the observations of single-cell approaches. For example, strains of *E. coli* tagged with the distinguishable cyan (*cfp*) and yellow (*yfp*) alleles of green fluorescent protein (*gfp*) in the chromosome showed their expressions fluctuated rapidly in time.[26] Such fluctuating responses of gene and protein expressions over time have also been observed in numerous other studies.[27] On the other hand, flow cytometry analyses have revealed cell-to-cell heterogeneity, Gaussian-like distributions for the abundance of a given protein per cell, in a clonal population of cells.[28]

The Origins of Biological Noise

The single-cell heterogeneity within cell populations, measured by transcription, phosphorylation, morphology and motility, arises from a combination of intrinsic and extrinsic elements. Increasingly, investigators recognize that the stochasticity in gene or protein expression is a result of two main sources of noise: (1) intrinsic or 'uncorrelated' noise; the random nature of biochemical reactions, e.g. due to low copy numbers of intracellular molecules in a Poisson process, and (2) extrinsic or 'correlated' noise; fluctuations in other cellular components or states that indirectly affect the expression of a specific gene or protein.

It has now been demonstrated, both computationally and experimentally, that stochasticity in mRNA and variability in protein expressions are not simply due to the effect of low copy number on a Poisson gene regulatory process, but rather due to the quantal or bursting nature of promoter activity. Moreover, by varying the rates of transcription and translation of a bacterial protein, it is now known that increasing transcriptional, and not translational, noise is responsible for the variability in reporter protein expressions.

Noise in Simple Gene Regulatory Network Is Crucial for Multimodal Decisions

For a long time, noise has been considered as an unwanted obstacle or nuisance in many disciplines, resulting in numerous efforts that focused on suppressing the cause and effect of noise. The usefulness of noise has gained valuable recognition across disciplines in the last few decades. In biology, however, it is only in the early part of the last decade that have we seen much interest and progress. This is mainly due to the lack of experimental techniques that were able to measure noise reliably.

In physics, it is well known that a small deviation in the initial condition could lead to diverse response of a deterministic process. This phenomenon, termed chaos, can be illustrated using a ski slope with bumps (Mogul skiing).[29] The gravitational force drives the skier to the finish point and the bumps create a landscape to develop error or noise that changes the lineage of each trajectory (attractor) to distinct pathways on different attempts. Hence, noise in a chaos process can be used to switch deterministic fates. For the cell differentiation process, how does a single cell diversify its lineage in time (Figure 1.3)? The clue may lie in biological noise on a chaos process.

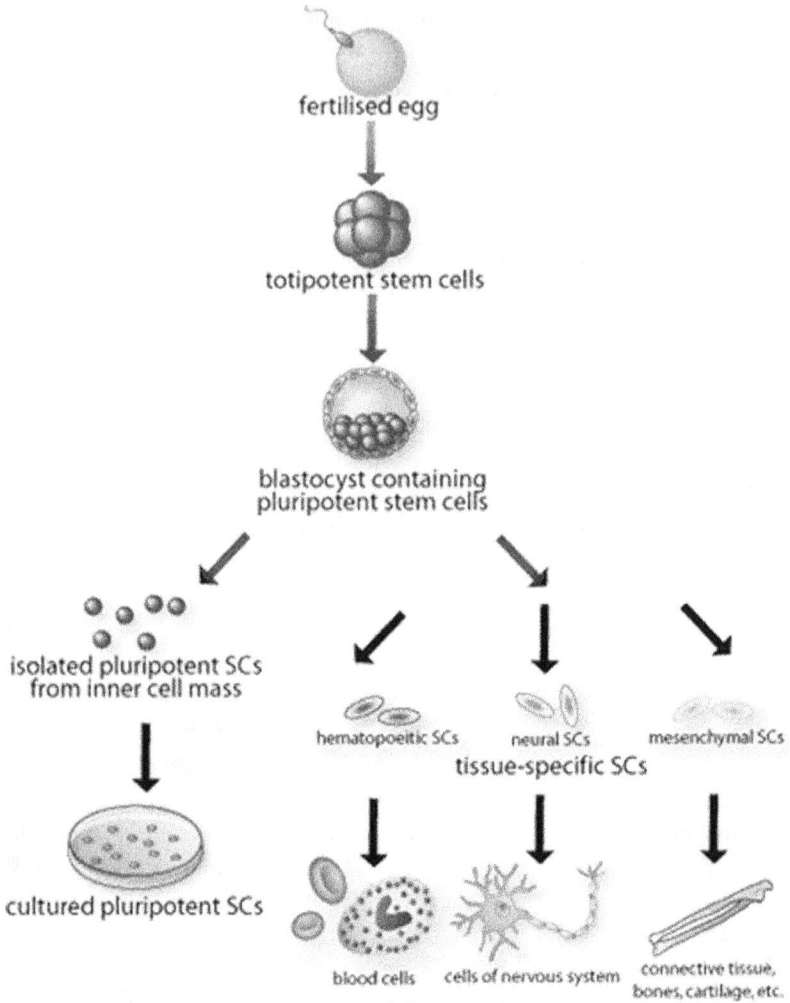

Figure 1.3: Schematic of mammalian cell differentiation process. Source: A. Chaudry, 'Stem Cell Bioengineering' (2004), at http://www.scq.ubc.ca/stem-cell -bioengineering.

Recent works in *Bacillus subtilis* have shown that biological noise or randomness in transcriptional machinery is crucial for controlling cell fate decision. Depending on the amount of nutrients available, *B. subtilis* survives in three modes: vegetative, competent and sporulative. A key molecule for switching states between vegetative and competence is the transcription factor ComK. In nutrient deficient conditions, *B. subtilis* survives in competent state by DNA uptake from the surrounding, facilitated by ComK which support the construction of DNA-binding and uptake apparatus. Lower concentration of ComK refers to vegetative state while higher concentration leads to competence, with an unknown threshold level switching between the states.[30] However, a population of *B. subtilis* in nutrient deficient conditions does not deterministically show all cells becoming competent. Rather, a mixture of vegetative, competent, and sporulated cells exists.

To understand the cell fate control mechanisms of *B. subtilis*, the expressions of comK mRNA was regulated with the construction of synthetic strains with low and high stochastic noise. *Rok* mutants, which showed high-levels of stochastic bursting, shifted the threshold of ComK concentration lower, thereby favouring competence. That is, although *B. subtilis* can exist in bistable states under nutrient-limited conditions, the phenotypic heterogeneity observed in the population can be controlled by the level of randomness in single cell dynamics. Another relevant work showed that the entry into and exit from the competent state occurred through an excitable core module with positive and negative feedback loops of ComK with ComS. Notably, the excitable dynamics of ComG and ComS generated by stochasticity was crucial for the regulation of *B. subtilis* competence.[31] Thus, living systems are able to harness stochasticity and variability to stay robust to adversity, i.e. change their cell fates using biological noise.[32]

Chaos in Biology

The preceding sections have indicated that biology should be appreciated from both microscopic (noisy) and macroscopic (deterministic) perspectives. For instance, although the control of stochasticity led to distinct cell fate in the competence of *B. subtilis*, the nuclear reprogramming by eggs and oocytes is deterministic, as it occurs with ordered and precise timing.[33] Furthermore, the early events of reprogramming mouse embryonic fibroblasts to pluripotency factor colonies also showed deterministic, rather than stochastic, steps.[34] In contrast to probability theories, these works suggest that well-defined deterministic processes must exist within the noisy cellular environment. Hence, the circumstances in which cells utilize noise (single cell) on deterministic (population) process remains unclear. As such, it is worthwhile to investigate further underlying mechanisms, apart from causal networks, which could utilize biological noise to change decisions.

In physics, the chaos theory appreciates the complexity of natural systems and attempts to formulate it using simple rules. Basically, chaos is referred to the existence of deterministic patterns from a highly non-linear system.[35] Its utility is recognized in systems as diverse as the weather, natural structures, the economy and the stock market. To illustrate a chaotic system, let us consider the Lorenz model, or a system of ordinary differential equations created to study atmospheric conditions.[36] Although simple, the model is highly sensitive to initial conditions.

When visualizing the result on a time-series scale, each simulation, within a sensitive parameter region, produced an unstable response.[37] However, when the overall results were transformed into phase-space plot (introduced by H. Poincaré[38]) clear patterns of the 'unstable' behaviour appeared with a core region, called the 'attractor', around which the solution tracks.[39] The stability of the atmospheric model simulations is revealed through the presence of the 'attractor' in phase-space plot, not in time-series, and this simple transition enabled the ability to see deterministic patterns from chaotic responses.

Today, there are several characteristics that a chaotic system is known to possess: (1) *Attractors*: when a system is perturbed, it evolves toward certain steady states. This response could be the return to the original state, or to a new state. Attractors can be fixed-point, limit-cycle, strange or chaotic. (2) *Sensitivity to initial conditions*: this is sometimes referred to as the butterfly effect. Any changes to the initial conditions can create very drastic changes to the final response. (3) *Fractals or self-similarity*: the scalable structure seen across the entire geometric construct. That is, as we go deeper and deeper into a chaos, we will realize that the smallest structure resembles that of the overall structure. (4) *Bifurcations*: the abrupt changes to system dynamics at specific point(s). (5) *Plasticity*: the ability to achieve multiple stable states to the system perturbations. For example, the self-organization of bacteria (individual) to form biofilms (population structure) when exposed to threats. (6) *Feedback loops*: information from one time point to the next could be fed back to the system for robustness to changing conditions. That is, system dynamics or trajectory is controllable due to feedback system. (7) *Phase locking*: the process when an individual diversion in a population could lead to the global shift in population behaviour. The way in which a shoal of tuna responds to attacks from whales is an example of this.

Attractors

Attractors are used to describe systems that evolve towards a specific state set by the initial conditions. To examine attractors in biology, high-dimensional gene expression analysis was performed for the differentiation of human promyelocytic leukaemia HL-60 cells into neutrophil by the action of two different reagents, dimethyl sulfoxide (DMSO) and all-trans-retinoic acid (atRA).[40] The time-course investigations showed the convergence of cell differentiation despite different initial transcriptome dynamics, demonstrating the existence of attractors for a preferred cell fate. Subsequently, another investigation sorting mouse haematopoietic cell lines into distinct concentration (low, medium and high) of stem-cell-surface marker *Sca-1* re-established the parental clonal heterogeneity profiles.[41] More recently, human cancer cell lines showed that sorting cells into distinct phenotypes (stem-like, basal and luminal) from parental population also returned the daughter subpopulations that resembled the parental population over time.[42] These data suggest the presence of strong equilibrium or attractor states in cell populations, whereby sorted (perturbed) subpopulations eventually return to their original state.

Self-Similarity and Patterns

In a human lung cancer study, a self-organizing map (SOM)-based statistical clustering technique was used to analyse the microarray data of types of tumours.[43] Remarkably, the SOM approach resulted in clear visual global gene expression patterns corresponding to distinct tumour subtypes of three randomly chosen individuals. Despite individual variability, the use of a statistical algorithm on a global scale allows one to detect the relationships between datasets, thereby revealing the presence of distinct cancer subtypes.

In another study, the temporal whole genome expression of Toll-like receptor (TLR) 4 stimulation of murine macrophages in wildtype, MyD88 knock-out (KO), TRIF-KO, and MyD88/TRIF Double KO (DKO) was investigated.[44] Temporal Pearson correlation analysis on the whole genome (over 20,000 genes/ORFs) was performed. To investigate whether 'scalability' or self-similarity of gene expression network exists, a random selection of 100 genes (repeated 30 times) was analysed, and compared with whole genome response for the 4 genotypes. The results showed that the temporal correlations of the random extraction followed the small, but reliable and monotonically related to time, departure from unit correlation pointing similarity to the whole genome response. This pattern between the whole and random grouping of genes, however, was absent for a selected 157 proinflammatory genes, especially in DKO where the monotonic response is almost absent. Thus, TLR4 induced gene expressions demonstrated self-similar structures found in chaos.

Overall, these investigations show biology as a system that resembles the characteristics of chaos theory, that is, the presence of order from what apparently seems like unpredictable or random information.

Sensitivity to Initial Conditions or Noise

Chaos is also characterized by its high sensitivity to initial conditions, resulting in diverging outcomes. Stochastic resonance (SR) is a phenomenon whereby the addition of a small periodic signal to a dynamical system that presents stochasticity amplifies the original signal, resulting in multistable states.[45] It follows that only when the signal-to-noise ratio reaches around a threshold level that such state changes occur. The concept of SR has been tested in biology. In neuroscience, the inter-spike intervals of periodically stimulated sensory neurons are multimodal. To uncover the mechanisms for the distribution behind the dynamics of bistable and excitability in neurons, Longtin[46] used a FitzHugh–Nagumo model, a simplified version of the Hodgkin–Huxley model,[47] and showed that noise on additive periodic forces are crucial in order to generate a multimodal distribution of inter-spike intervals of neurons firing to a specific phase of the input signal. Although the model could not fully interpret the peak heights of the spikes, nevertheless, it demonstrated that neuron-firing activity is positively governed by a simple non-linear model, with SR phenomenon controlling the multimodal response.

The sensitivity to noise for stem cell differentiation was recently investigated with the aid of the dynamical computational model.[48] To understand the fundamental mechanisms for stemness and plasticity in multipotent stem cell differentiation, dynamic gene regulatory network (GRN) models using simple non-linear equations that considered cell-to-cell interaction and the effects of heterogeneity was developed. Notably, among more than a hundred million GRNs tested, GRNs that produce oscillatory protein expression dynamics and whose synchronies lost between cells were shown to be crucial for stem cell differentiation and proliferation. This result, which is in agreement with experimental observations,[49] indicates that the complex process of cell differentiation may be regulated by GRNs governed by simple non-linear reactions that produce multistable outcomes (attractors) when biological noise in terms of stochasticity or heterogeneity is added.

In summary, the examples presented so far demonstrate that biology possesses several characteristics of a chaotic system that can operate in a far-from-equilibrium state: (1) biological noise can destabilize original steady states and induce transitions for new regularity in a time dependent manner,[50] e.g. in the bifurcations of ComK expressions,[51] (2) cell differentiation and synaptic transmission processes have been shown to cross the barrier between stable attractors by noise,[52] (3) the robustness of several key processes to biochemical variation as witnessed in bacterial chemotaxis,[53] (4) bimodality behaviour in yeast required noise on a feedback mechanism,[54] (5) the phase locking of bacteria in biofilm formation,[55] and (6) the suppression of bursting mRNA into ordered decision in yeast transcriptional system.[56] These properties, collectively, enable biology to attain multiple but finite number of attractor states.

Conclusion

At cell population level, biological pathways are guided by deterministic response waves. These waves can be modelled using simple mass action equations. From cell signalling examples, it was shown that network topology, rather than reaction parameter values, is crucial for interpreting signaling dynamics. That is, we observe simple statistical laws, such as linear response and power-law, which are important to shape global response. However, from single-cell and single-molecule investigations, stochastic molecular interactions and cell-to-cell variability are shown to be key to initiate multimodal decision. Biology also resembles chaotic systems through attractor states, self-similarity and sensitivity to noise.

Biology, thus, like any other complex system, possesses both microscopic (single cell) and macroscopic (population average cell) dynamics.[57] Combining cell population and single cell behaviours suggest that biology is regulated by deterministic governing equations, and is sensitive to parameter variations (noise) over a specific range, as witnessed in other disciplines such as physics, chemistry and engineering.

I believe this chapter has provided key awareness of the dynamic behaviour of living cells. This book contains other insightful chapters that describe various alternative ways in dealing or approaching biological complexities. The recent advent of multidisciplinary techniques is also helping us to gain further understanding on the evolutionary and adaptive properties of living systems.

2 NEWS FROM THE 'TWILIGHT ZONE': PROTEIN MOLECULES BETWEEN THE CRYSTAL AND THE WATCH

Alessandro Giuliani

Setting the Scene

Some Very General Facts about Proteins

Proteins occupy a very peculiar niche between chemistry and biology.[1] Proteins are large, complex molecules that any polymer chemist would have difficulty in modelling. They are long (from 50 to 10,000 elements) strings made by non-periodic sequences of 20 monomeric chemical species (aminoacids) kept together by a covalent link (peptide bond). We refer to the linear arrangement of the aminoacids along the sequence as the 'primary structure' of the protein molecule. Primary structure is folded in solution so to give rise to the three-dimensional architecture of the protein: nearby aminoacids acquire three characteristic mutual arrangements: alpha-helix, beta-sheet and random-coil.[2] The entire sequence then folds 'on itself' establishing long-range contacts between aminoacids very distant along the sequence, giving rise to the tertiary structure. Eventually some proteins, thanks to non-covalent contacts between different sequences, acquire a 'quaternary structure' generated by the juxtaposition of different folds. On the biological side, although no single protein can be considered as alive, it does not take many of them (plus a bit of nucleic acid) before life-like behaviour begins to emerge. For example, some of the smallest viruses, such as HIV, lying on the borderline of life, are endowed with only 10 different types of proteins.[3]

Moreover each protein molecule exerts a very specific 'biological function' strictly dependent upon its structural arrangement. To further enrich the picture, it is worth noting many proteins exert their physiological role by assuming many different folds depending upon their chemico-physical microenvironment. The biochemists refer to these molecular systems as 'natively-unfolded' proteins that are able to adapt their configurations 'sensing' their microenvironment.[4]

Proteins are synthetized from DNA strings (genes) that, by the action of genetic code, translate a given nucleotide sequence into the corresponding protein primary structure.[5]

Scientists refer to this statement as the 'molecular biology dogma' that is often referred as 'one gene – one protein'. **In the last decades the discovery protein species largely outnumber (of around an order of magnitude) the number of genes has falsified the dogma.**[6] Many 'periphery-driven' processes as alternative splicing or post-translational tissue-specific modifications are at the basis of the lack of one-to-one correspondence between genes and proteins.[7]

Protein Molecules Priorities

While artificial polymers are insoluble and precipitate into amorphous multi-molecular aggregates, proteins are water-soluble and each molecule maintains its individuality in solution. The task of being water soluble while maintaining the structural specificity necessary for a physiologically motivated activity is not easy, and only a relative minority of linear amino acid arrangements can actually accomplish this goal.[8]

The difficulty of the task stems from the huge dimension of proteins with respect to other molecules and to the need of establishing very delicate and intermingled relations with water molecules. There are no clear boundaries between the internal and external (exposed to solvent) parts of protein molecules: proteins behave as sponges presenting hydrated cavities where water molecules accommodate into the protein molecule 'cage'.[9] This 'internal water' plays a crucial role in protein activity.[10] The reach of a given threshold of protein hydration is a necessary pre-requisite for metabolism. Many 'suspended' forms of life (e.g. seeds, spores) start an active metabolism following the reaching of a threshold of protein molecules hydration allowing the generation of a continuous 'bound-water' network.[11]

From what said before, we can safely state that, if you are a protein, your main priority is 'to be able to swim' or, more rigorously: to be soluble in water without losing your native configuration.

The concept of structure resonates with something very stable and immutable. The case of protein three-dimensional structures forces us to a different thinking. A protein, in order to be soluble and to exert its physiological role, must have meaningful dynamics, in other words, the structure must be highly flexible. The free energy difference between the native (the protein arranged in the physiologically efficient way) and denatured conformations is small, this implies folds are only marginally stable. This allows for the necessary flexibility, letting the molecule adopt slightly different conformations.[12]

Because of the perturbations arising from molecular collisions due to thermal motion, each fold is continually subject to conformational rearrangements.[13] A given fold is able to regain its native conformation, being a natural free-energy minimum, the native conformation acts as a dynamical 'attractor' drawing all the parts of the fold back to the native configuration. Just as a ball in a bowl always ends up at the bottom, a protein molecule is capable to recover its native configuration dissipating the intervening transients.

This is a second basic priority for a 'well-behaving' protein: to have a structural arrangement that guarantees at the same time flexibility and robustness. Last (but not least) a protein sequence must attain its native fold in reasonably short times ranging from milliseconds to minutes.

A purely random, diffusion driven, search for the optimal configuration will last tenyears for a medium-size protein (this is the basis of the so-called 'Levinthal paradox', named after Cyrus Levinthal, who first stated the problem).[14]

This means the protein molecule must have 'embedded' somewhere not only the information for producing a soluble and stable configuration but even the recipe to reach this configuration along a very ordered, non-random strategy.

Until now we have not approached the problem of the 'protein biological role'; all the above features pertain to the chemico-physical realm, and are 'givens of physics',[15] in other words we are still in the field of 'Logos' where forms are the consequence of general natural laws.

In a fundamental paper published in 1994 entitled 'Biomolecules: Where Physics of Simplicity and ComplexityMeet' by Hans Frauenfelder and Peter Wolynes[16] focused on the need to have physics principles of 'simple' systems (like atoms) cooperatively interacting to produce macroscopic principles describing the complex features of protein molecules. While we do have an accurate knowledge of potentials (hydrophobic interactions, hydrogen bonding, size constraints, etc.) acting at microscopic levels, the 'mesoscopic' principles needed to accurately predict the above described protein features remain essentially unknown. Notwithstanding that, such principles must exist (and they are a powerful driver for physical chemistry research). In other words, as we accept valence rules and energy requirements governing organic chemistry are 'already there' and hold even for not-yet-synthetized molecules, we are forced to think that the subtle balance between stability and flexibility characterizing proteins can be explained by physics-based considerations without asking for the selection (function-based) forces acting in biological evolution.

Only after having solved the above structural problems, a protein can start to think about what she can do for the organism (whether it is a bacterium or a human being) hosting her (I prefer to think of proteins as female). Here we leave the 'Logos' (or in any case a lawful way of doing science) and enter the 'Bios', that is, a world ruled by functional optimization 'imposed' by higher hierarchical levels. The heart features are 'functionally optimal' not for the heart in itself but for the entire organism, and the same holds true for an enzymatically efficient protein with respect to a mutated protein with less efficient catalytic activity. The 'biological priorities' of protein molecules make us shift to a new model of organic form: that of the machine or artefact. While solubility, stability, flexibility and folding rate tell us of forms as 'built-in' lawful givens of physics (like crystals, clouds or hydrodynamic vortices), the emphasis on biological function subordinates the protein form to its specific biological task. In the same way the configuration of a watch cannot be understood without referring to the task of measuring time, the purely functionalist approach makes the 'Bios' departing from 'Logos' (Crystal) to go toward the 'Techno' (Watch'). In the following section I will discuss why protein molecules (even if very well suited for their physiological function) are much more similar to 'Crystals' than to 'Watches'.

Smart and Sensible Crystals

Some Facts

We were left with the task of looking for the signatures of 'functional' and 'structural' (in the broad sense) optimization in actual protein molecules and eventually trying and estimating the relative weight of the two modes. As first, it is important to stress the fact that the protein molecules we deal with are all coming from biological organisms: *de novo* synthesis of protein molecules in abiotic conditions is extremely hard and the (very few) purely synthetic constructs are much simpler than the simplest natural protein. Chemists produce synthetic proteins relying on natural templates and inserting various modifications by genetic engineering techniques.[17]

We refer to protein molecules by their main biological activity, thus cytochrome c oxidase (COX) is the last enzyme in the respiratory electron transport chainof mitochondria (or bacteria) located in the mitochondrial (or bacterial) membrane. It receives an electron from each of four cytochrome c molecules, and transfers them to one oxygen molecule, converting molecular oxygento two molecules of water. In the process, it binds four protons from the inner aqueous phase to make water, and in addition translocates four protons across the membrane, helping to establish a transmembrane difference of proton electrochemical potentialthat the ATP synthasethen uses to synthesize ATP. The summary reaction catalyzed by this enzyme is:

$$4 \, Fe^{2+}\text{-cytochrome } c + 8 \, H^+_{in} + O_2 \rightarrow 4 \, Fe^{3+}\text{-cytochrome } c + 2 \, H_2O + 4 \, H^+_{out.}$$

The very same reaction happens in all the organisms (from bacteria to horses and humans) where the cytochrome c oxidase (COX) is present and the COX coming from different organisms are practically superimposable in terms of structure. This identity derives from the fact that structural requirements of COX are very demanding: the protein interacts very specifically with other proteins and with the membrane and the fulfillment of these constraints permits only a very limited range of 'allowed' mutations. Defects involving genetic mutations altering COX functionality or structure can result in severe, often fatal metabolic disorders.[18]

This situation allows us to grasp the first aspect of the 'natural laws' vs 'biological evolution' combined action on proteins: the complex functional role of COX strictly determines the protein sequence (and consequently structure), this drastically shrinks the allowed mutation range. From the point of view of a COX molecule, there is no relevant difference to being inside a bacterium or a dog: the signatures of biological evolution are practically absent in protein sequence. COX molecules from different organisms are practically identical. The need to perform a given physiological task imposes very hard constraints on the protein architecture and even relatively tiny variations are filtered out by natural selection.

With very few exceptions, protein molecules do not work by alone but by interacting with other protein species. This fact makes interacting proteins co-evolve: the need to maintain an efficient between protein interaction between two A and B interacting protein species drastically limits the space of 'allowed mutations' of the two species, so making the mutational spaces of A and B to correlate.[19]

The computation of the correlation between the mutational pattern of two proteins is routinely used to infer protein–protein interaction: a correlation threshold around $r = 0.70$–0.80 with r being the Pearson correlation coefficient between the phylogenetic trees built from A and B respectively allows for a very efficient interaction recognition.[20] The same phenomenon appears even considering a single protein species: the need to accommodate a given amino-acid mutation into a coherent system that must be 'stable' (i.e. a chemico-physically optimal molecule) makes the observed mutations in a specific amino-acid to go together with mutations in other parts of the molecule so to globally reach a stable protein.[21]

This corresponds to a sort of 'reversed direction' filter with respect to the one discussed in the case of COX: the need of having a stable, soluble and in general well-behaved chemical entity drastically limits the effect of random mutations, the starting material of biological evolution. This limitation (or better 'canalization' as evident by the induced correlation between different molecules evolution) happens before any 'organism level phenotype' is evident. This gives a proof-of-concept to the Stuart Kauffmann statement of a greater role played by physical constraints with respect to the classical neo-Darwinian model that assigns a prominent role to 'chance + selection' evolution mechanisms over physical constraints.[22]

From our point of view, it is worth noting the natural superposition and interaction of 'Bios' and 'Logos' in the 'twilight zone' between chemistry and biology.

If Someone Is Smart, They Are Smart in Any Occasion

Protein Data Bank (PDB),[23] the biggest repository of protein structures, contains 105,128 structures (as of 6 May 2014). These protein structures corresponds to around 1,000 distinct folds.[24] This is not unexpected: if there are some 'optimality principles' at work in the determination of protein structures, we do expect that the 'allowed forms' are much less than the theoretically possible forms. What is very interesting is that the proteins inside the same fold class do not share any evident biological function commonality.[25] This is another proof-of-concept of the fact that 'Bios' acts on 'givens of physics' that were globally optimized in terms of non-biological features, the 'biological optimality' deals with 'fine tuning' on already 'physically optimized' objects.

Adopting a topological view, we can go further in depth into the physical optimality criteria of proteins.[26]

The full-rank information about a given protein native structure coincides with the three-dimensional coordinates of all its atoms. Given the constraints existing between the mutual position of the different atoms, we can reduce the burden of data keeping intact the initial information by limiting ourselves to the coordinates of alpha-carbons (the carbon atoms adjacent to the peptide bond). The knowledge of the alpha-carbons coordinates is sufficient to generate the usual ribbon structures of proteins like the one in Figure 2.1.

We can go further in this reduction: from basic physical chemistry we know two amino-acids can be considered to be in contact if the mutual distance between their alpha-carbons in the 3D Euclidean space is lower than 8 Angstroms. On the other hand, two consecutive amino-acids along the sequence are 4 Angstrom distant from each other in the 3D space. The two above facts imply that the 'non-trivial' (i.e. not obliged by the sequence proximity) spatial contacts between amino-acid residues in a protein molecule are located in the range between 4 and 8 Angstroms. This allows us to generate a contact map (corresponding to a graph) from a protein 3D structure considering only the yes/no information about the existence of a contact between each pair of amino-acid residues.[27]

Considering only the contact map information allows for a better clarification of protein architectural principles with respect to the full-rank architecture[28] and shows the presence of topological principles shared by all the protein molecules.

A very intriguing general (common to all the protein molecules) principles is the minimization of average shortest path (asp) of the protein graphs. A shortest path between two nodes (i) and (j) in a graph is the minimal number of edges (contacts) to be traversed for going from (i) to (j) (and vice versa). The average length of the shortest paths over all the pairs of residues is the asp of a graph (also called characteristic length of the graph), a graph architecture that minimizes asp at a given contact density is an optimal architecture in terms of wiring and efficiency of communication among its parts. Protein graphs are optimal as for asp minimization, this feature derives by purely chemico-physical constraints and is achieved by a 'smart' location of long-range contacts (contacts between residues far away in sequence, see Figure 2.1).

The 'smartness' of long-range contacts positioning derives from energy minimization during the folding process (given of physics) but ends up as a very desirable biological consequence: the possibility of giving rise to efficient allosteric effect.[29]

The protein reacts very rapidly to the binding of an effector in a given part of the molecule thanks to the fact the signal spreads very rapidly across the asp-minimized graph (the definition of allostery is exactly the change of configuration of a site distant from the event of binding in consequence of the binding event).

The abstract character of the graph representation allows us to purposely adopt the protein graph architectures in any man-made artefact (be it a subway line, a firm organization, a sensor web). In this case we step into the 'Techno' realm by means of a bio-inspired solution. The important point to keep in mind is that biology (allosteric effect) and technology (artificial network architectures) derives 'for free' this smart tip from a natural system (protein folding) that did not 'purposely' followed this goal. In other words, from a crystal and not from a watch.

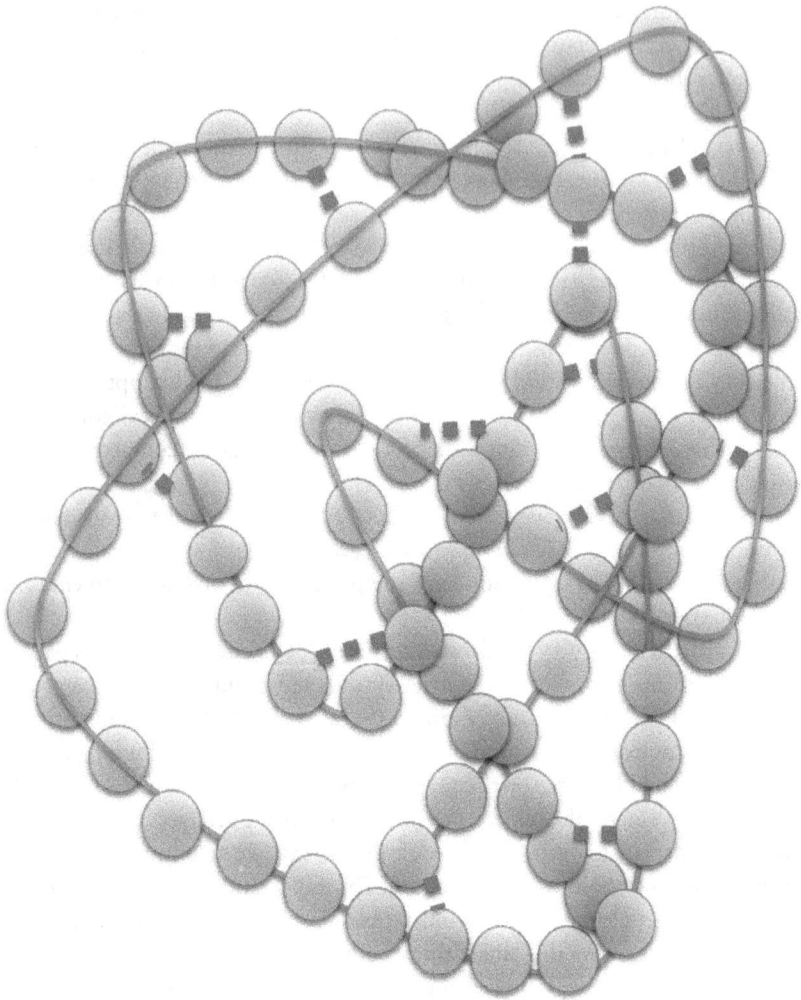

intramolecular
noncovalent interaction

Figure 2.1: Minimal model of a protein molecule. The bead of pearls corresponds to a proteins folded into its native structure (pearls = amino-acids). Long-range contacts are labelled by dashed lines; the disposition of long-range contacts in the protein is driven by asp minimization (see text).

Conclusions

I hope to have conveyed the concept of how the analysis of protein structure allows us to go in-depth into the mutual relations between 'Bios', 'Techno' and 'Logos', so giving scholars a perfect playground to test their ideas in practice (more than 100.000 protein structures are waiting out there). Going back to the initial question of the relative position of protein molecules (these molecules, do not forget, are located exactly where life begins and are the direct product of the genes) in the range between the crystal and the watch, the answer is: 'proteins are much more similar to crystals!' The intriguing point is that they are 'smart and sensible crystals' that *can be used as watches*.

For the above reason they can be studied using all the three 'Bios' 'Techno' and 'Logos' attitudes. The opposition between crystal (structuralist) and watch (functionalist) views has an impact in the general philosophy of biology, in this respect the comment reported by Dentonand colleagues[30] is enlightening:

> We note in passing that there is some irony in the fact that in adopting the metaphor of the artifact or machine, the Darwinists had adopted the same metaphor as their creationist opponents. For both creationists and Darwinists, life's order is contingent and artifactual, like the order of a machine, like the order of the Paleys' watch[...] For creationists it was God who had controlled life's contingent order, for Darwinists (Dawkins) it was a 'Blind watchmaker' who relied on time and chance.

The structuralist view allows for a new enchantment for life that is located in a continuum from the birth of the universe to today's still acting physical principles, and far from 'just-so-stories' and *post hoc* explanations.[31] The engineer can contemplate with an open mind 'nature-as-it-is' without any anxiety to control it or immediately asking 'what-it-is-for' at any moment: smart solutions can come from everywhere.

3 LIMITS TO DETERMINISTIC-LINEAR CAUSALITY IN BIOMEDICINE: EFFECTS OF STOCHASTICITY AND NON-LINEARITY IN MOLECULAR NETWORKS

Sui Huang

In this chapter we examine from the perspective of the theory of complex systems the notion of proximate causation in medicine: how can a microscopic event, embodied by the presence of a biomolecule or by a molecular process, explain a 'macroscopic biological observable', that is, a phenotypic change, such as a disease? More broadly in biology such causal associations are at the heart of the problem of the *genotype-to-phenotype mapping:*[1] How does a gene activity (or that of its allelic variant) explain a phenotype? Obviously this question is of central importance in our intellectual quest to understand how genes control the phenotype. But it is also of eminent practical significance in medicine: drug treatment of a disease is in essence the modulation of a phenotype by intervening with molecular processes with the goal of steering it from a disease state to a healthy state.[2] In the past decade, driven by medical relevance, the old question of mapping between the microscopic world of biomolecules (typically, the gene products) and the macroscopic world of phenotypic behaviour has moved under the new light of systems biology. This youngest discipline in biology arrived around 2000 in the wake of the technological revolution brought by genomic technologies. It seeks a new type of insight by first exhaustively collecting, categorizing and characterizing all the molecular and cellular parts of living organisms, which now appears achievable with the arrival of massively parallel 'omics' technologies. This is followed by the functional analysis to determine the interactions between these component parts, which in turn establish molecular networks, most notably gene regulatory networks(GRN), through which individual gene loci influence each other's activity. Mathematical models can then be employed to analyse how the interactions of these component parts produce the collective behaviour of biomolecules that is manifest as the integrated or 'emergent phenotype'.[3]

Some phenotypes are not actually emergent and can readily be explained in a straightforward manner from the genotype. But such cases of a 1:1 mapping between gene and trait, or mutation and disease, are exceptions. They imply a linear, deterministic cause–effect relationship: '*A* (a genotype) causes *B* (a phenotypic trait)'. Examples of such linear-deterministic causality include monogenic inherited diseases, such as sickle cell anemia or an inborn disease of metabolism where the biochemical defect in the protein is predicted by the mutation in the DNA and explains the phenotype. Such cases are islands of linear causation in a sea of non-linear relationships. In the course of this chapter the precise meaning of the terms '*linear*' and '*non-linear*', which we so far use in a loose manner according to common usage, will become clear. Only linear relationships permit the articulation of a straightforward and plausible causal relationship the sense of everyday usage: an explanation of effect *B* by the presence of cause *A* without use of mathematical formalism. For now, 'non-linear' should stand for complex, convoluted relationships that defy the '*A* causes *B*' type of explanations. One goal of a systems or integrative approach is to understand the emergence of integrated organismal behaviour, that is, ultimately to understand why the whole is more than the sum of its parts for which linear causation is not sufficient.

The traditional epistemic habit of molecular biology, collecting the parts list and making sense of them by 'hand-waving' and search for linear causal explanations for phenotypes, is epitomized by the ubiquitous 'arrow–arrow' schemes (Figure 3.1, top): $A{\rightarrow}B$, which stands for '*A* causes *B*', where *A* is typically a biological molecule. In modern biology parlance, A is said to be 'upstream' of *B*. Here a scheme of causation and the actual physical (molecular) embedment, the pathway, are congruent. This is only permissive in the simple case of the 'islands of linearity', as we will see. The more such schemes of causation, all embodied by molecular entities, we know, so went the thinking for a while, the more we could explain. While this culture of thought has been broadly (and vaguely) criticized as 'reductionism', there is now high awareness that the new massively-parallel analysis of many molecular pathways ('$A{\rightarrow}B{\rightarrow}C{\rightarrow}...$') must avoid becoming simply the extension of traditional pathway analysis, so to speak, massively-parallel reductionism. All-encompassing, systems-wide analyses are necessary but we need to complement this '*entirety of analysis*' with the '*analysis of entirety*'.[4]

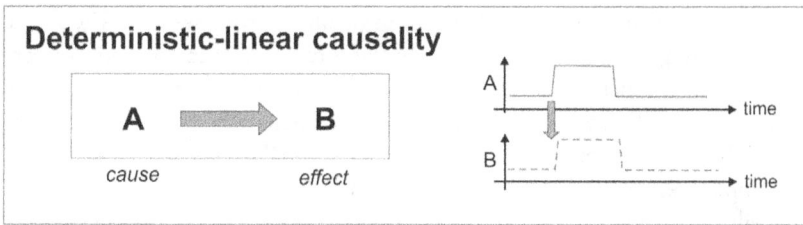

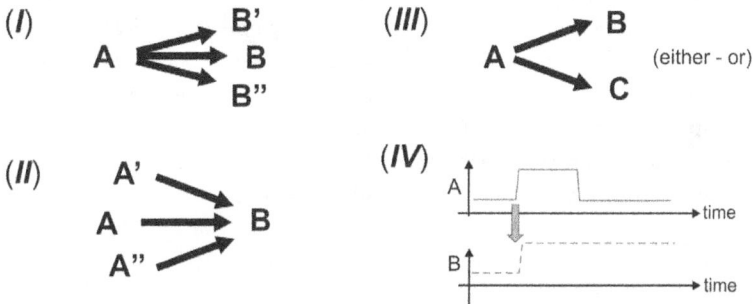

Figure 3.1: Schematic overview of problems of linear-deterministic cause–effect relationships in complex systems. The intuitive, common-sense deterministic, linear causality (box) and the four types of departure from it, numbered as discussed in the text *(I)–(IV)*. 'A' is the cause, and 'B' and 'C' are the effects.

This will require some technical knowledge about formal tools for describing, hence, understanding complex systems. This chapter seeks to offer central technical concepts to achieve this. It is far from a discussion about the philosophy of causality in biology, which is a domain of research in itself, nor is it about the dichotomy between proximal and ultimate cause in biology,[5] a matter that continues to engender vivid debate, notably in the explanation of complex living systems. Instead, this chapter aims at presenting a set of fundamental phenomena in the behaviour of living systems that can only be described using the principles of non-linear dynamical systems whose relevance in biology has attracted increasing scrutiny in the past years in systems biology but have yet to be embraced by 'mainstream' biologists. Second, in addition to non-linear behaviours we will also introduce a more fundamental aspect of dynamical systems: stochasticity, or non-deterministic behaviour, which also challenges the common notion of causality in biology which tacitly assumes deterministic cause–effect relationships.

We will present concepts that will serve both experimental biologists habituated to reduce phenotypes to molecular pathways as well as epistemologists of life sciences seeking to study explanatory principles in living systems as 'tools for thought'. They hopefully will help, even without detailed mathematical treatment, to comprehend phenomena that abound in living systems but appear counterintuitive because they defy the intuition of deterministic linear causation.

We hope that through the use of these conceptual tools, many observations that appear paradoxical or at least counterintuitive can be naturally and satisfactorily accounted for. Of course, a behaviour that is puzzling to one will be expected by another person because of distinct preconceptions of linear relationships and distinct degrees of familiarity with dynamical systems. For instance, the finding that a cellular response persists long after a perturbation that triggered it has subsided is counterintuitive to many molecular biologists but not so to physicists. We do not suggest that non-linear and stochastic dynamics will reconcile all paradoxes but that at the very least they will come as a handy tool to fill a gap of understanding that is not due to an inherent evasiveness of biology but that is kept open rather because of the failure to study elementary biological phenomena through the optics of dynamical systems.

Where the Problem Lies

We ask the simple question: what are the conceptual obstacles encountered when assigning to a gene or its encoded protein a simple causative role in producing a particular phenotype? What is behind the failure of establishing a linear causality between a biomolecule and a phenotype?

Traditional Linear, Deterministic Causality

Assume, for sake of argumentation, the simple case: $A{\rightarrow}B$, where A is a gene activity (the gene is present and its encoded protein expressed) and B the activity of its target gene, or a phenotype triggered by A. We refer here to such a relationship as **deterministic, linear causality**: 'presence of activity of A (implicitly always) causes phenotype B' (Figure 3.1, top). If one were now to use this '$A{\rightarrow}B$' scheme to explain phenomena in living systems, a common practice in molecular biology, one would blame the inability to explain a particular phenotype with $A{\rightarrow}B$, or any paradoxical observation inconsistent with $A{\rightarrow}B$ (e.g. presence of A but absence of B) simply to the lack of knowledge of additional component parts that might play a role. For instance, failure to suppress B following deletion of A is typically explained by the assumption of 'redundancy' – the presence of alternative pathways that causes B. One is of course also aware of conditionality: if suppressing gene A (as in 'gene knockout experiment') eliminates phenotype B, then A is said to be *necessary* for B, otherwise it is not. By contrast, if overexpression of A (always) causes phenotype B, A is said to be *sufficient* for effect B.

It is common practice of genetics and molecular biology to assume that, when a necessary and sufficient role of *A* in causing *B* is demonstrated, this establishes a strong causal relationship.

But in reality it will not be as simple if one considers that *A* and *B* are members of not a chain of causation but of a network of causal relationships that involves non-linear and non-deterministic interactions to be discussed here.

Departure from Linear Causality

To overcome the limits of straightforward causality (see Figure 3.1) we will need to go beyond classifying causes as necessary, sufficient or both. The following recurring scenarios in which the relationship between cause *A* (activation of gene *A*) and effect *B* (its phenotypic) depart from the simple notion of causality of geneticists and molecular biologists will be the focus of this discussion (Figure 3.1, bottom):

(*I*) *A* can cause *B*, but does not always do so. More often it causes *B'* which is similar but no identical to *B*. Moreover, *B'* varies each time *A* is activated, fluctuating around *B*. This constitutes 'weak causality', and makes precise prediction impossible but allows for accurate statistical predictions ('on average *A* causes *B'*).

(*II*) *A* causes *B*, but so does *A'* and *A"*, etc. – a set of causes, e.g. activation of genes *A'*, *A"*, etc. will result in the production of phenotype *B*. Thus the causal mechanisms behind *B* is not unique.

(*III*) *A* can causes *B*, but not always – only in a fraction of times. At other times *A* causes *C*. Unlike the above case (I) the various outcomes are *discretely* distinct, and there is a finite number of outcomes, typically two (*B* and *C*) – in which case a *binary* decision making between alternate choices is involved. As in (*I*) the outcome is indeterminate.

(*IV*) *A*, activated only transiently, cause *B*, but upon deactivation, *B* persists. Here a transitory cause has lasting effect, hence a memory is involved

The counterintuitive departure from the linear, deterministic 'straightforward' *A→B* causation, of the above scenarios that are frequently encountered in biology is not simply due to lack of knowledge of certain molecular facts (unknown interactions), an assumption that would rescue the concept of linear deterministic causation (for instance for (*I*) and (*III*) one may assume a graph with many arrows from *A* to *B*, *B'* *B'* or to *C*). Such attempts at rescuing the notion of a 'straightforward' *A→B* causation by adapting the arrow–arrow causation graphs accordingly to account for counterintuitive phenotypic behaviours are not unlike the epicycles of Ptolemy.

Therefore, in the next two sections we will discuss the two principles, stochasticity ('Gene Expression Noise and Stachasticity in Biological Systems') and non-linearity ('Non-Linearity in Gene Regulatory Networks') with sufficient detail to provide a conceptual framework that will help to comprehend the principles underlying the four non-intuitive departures from the notion of linear, deterministic causation listed above and how they still can be related to the molecular interaction maps that were used as schemes of networks of causal networks.

Gene Expression Noise and Stochasticity in Biological Systems

The most prosaic culprit for the non-intuitive deviation from the elementary notion of the $A{\to}B$ causation in molecular biology is the departure from the deterministic relationship implied by such arrow–arrow schemes. The chief source of non-deterministic dynamics is what has become known as *gene expression noise* cell.

History of Stochasticity in Cell Biology and Early Concepts

'Gene expression noise' or more generally, 'molecular noise' or 'cellular noise' describes the apparently random fluctuations of the intracellular abundance of a given protein (and other biomolecules). In the case of proteins it is the result of the series of stochastic (= random) events that are part of the processes in which a gene locus is activated, its gene transcribed and the messenger RNA translated into a protein.[6] Stochastic fluctuations originate from chemical reactions in which the participating molecules are locally present at small copy numbers within the cell. This is particularly pronounced in gene expression: the reaction is initiated by one to two target DNA binding sites (promoter sequences) for a given gene locus and a few hundreds to thousands of specific transcription factors that bind to it. As a consequence of the small numbers of players involved, the random length of time intervals between the collisions of molecules cannot 'average out'. Collectively, the random molecular processes would result in temporal fluctuations of the abundance of the observed end product X, notably in the form of 'bursts' of production because the assembly of the production apparatus (transcription initiation complexes, etc.) is relatively unlikely compared to subsequent rounds of synthesis of the transcript. The visible result is a considerable cell-to-cell variability of the cellular abundance of that protein species X within a cell population at a given time point.

While this cartoonish picture is in principle correct, the biochemistry of gene expression is in reality much more complex than just the two-step sequence of transcription and translation; many more noisy processes are involved, such as opening of chromatin, transport of mRNA out of the nucleus, splicing of introns, etc., each contributing fluctuations at varying time scales, ultimately giving rise to the noisy fluctuations of protein levels. Gene expression noise has become popular to study in the late 1990s when the technology of **recombinant fluorescence**

protein afforded a convenient means to visualize and quantify the abundance of a protein in individual live cells within a large population of apparently uniform cells.[7] However, the notion that stochastic events play a role in biological processes within cells had already been proposed by Kupiec in the 1980s.[8]

Measurements of Stochasticity in Cell Biology

The measurement in **flow cytometry** of fluorescently labelled cell surface proteins, or of fluorescent proteins, in a suspension of cells at single-cell resolution which produces histograms of the statistical distribution of cellular abundance of specific proteins (= 'bell shape' histograms of the number of individual cells that have with a given cellular protein abundance in 10,000s of cells, Figure 3.2) came into routine usage in the 1980s. It was, however, used mainly for detection of distinct subpopulations of cells and the variance of cellular protein abundance within an apparently 'pure' subpopulation disregarded. Thus, the measurement result was interpreted as a deterministic quantity. Its value in providing insight in stochasticity, manifest in the invariably broad statistical distributions, had initially been ignored because the dispersion of values between the cells were considered technical noise (measurement errors).

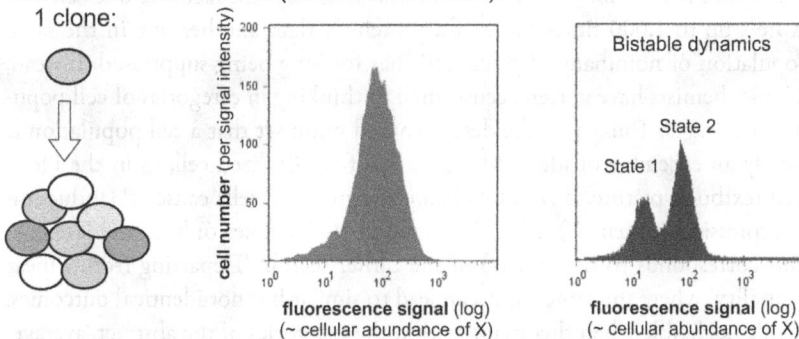

Figure 3.2: Cell-to-cell variability due to stochasticity of gene expression and two examples of flow cytometry histograms. Note the large dispersion of the signal value (x-axis) which represents proportionally the abundance of a specific protein in an individual cell.

Interpretation of the spread of the distribution as manifestation of gene expression noise however must consider that the time scale of fluctuations in the individual cells is lost in the histograms which reflect snapshots of the entire cell population at one time point. The significance of the outliers in the 'tails' of the distribution depend on whether (*i*) they are frequently visited by individual cells when they fluctuate and are short lived – in which case these cells are statistical flukes – or (*ii*) they are visited very rarely but in turn, are long lived and return slowly to the bulk around the mean of the population. The former case reflects *ergodic behaviour*: here a snapshot of the entire population yields the same information as taking many measurements of one rapidly fluctuating cell over time and assembling the values to a histogram. Flow cytometry histograms (Figure 3.2) of cell populations cannot distinguish between these two possibilities and initially were by default interpreted as the ergodic case. But mammalian cells appear to exhibit long-enduring outliers with respect to individual fluctuating properties which confer biological functionality[9] – hence the notion of noisy, meaningless 'fluctuation' and time-independent type-diversity within a population becomes blurred.

While flow cytometry measurements destroy the cells measured or at least forget which cell is which after one measurement, more cumbersome longitudinal monitoring of the very same cell under the microscope, akin to filming the blossoming of a flower with a time lapse camera, can detect the shape of the fluctuations and hence distinguish between the ergodic or non-ergodic case. However, special methods such as fluorescent protein tagging of individual proteins of interest (which requires genetic manipulation of the cell line) must be used for such non-destructive monitoring of the same cell over time.[10]

Implications for the departure from linear causality: the fact that one cell may express up to 1,000 times more of a protein X than another one in the same population of nominally identical cells has for long being suppressed. Instead, cell biochemists have gotten accustomed to thinking in categories of cell population *averages*. Thus, with the deeply rooted mind-set that a cell population is merely an ensemble of identical replicates of an idealized cell, as in the idealized textbook picture, if A (a biochemical signal to a cell) causes B (induction of expression protein X), then the causality $A{\rightarrow}B$ applies only to the 'average'. This corresponds to the case (I) of the earlier section 'Departing from Linear Causality', where the same cause can lead to similar but not identical outcomes. While negligible when theorizing within the categories of the abstract 'average' cell, it becomes relevant in physical reality because the fate of *individual* cells could under specific circumstances be amplified and determine the outcome of an entire cell population. This is particular relevant in the non-ergodic case with enduring outliers. Then, 'averaging' does not rescue deterministic causality and the same causes may have different effects – the case (III) which is explained later.

Randomness and Deterministic Causality in Macroscopic Biology

Randomness is profoundly at odds with our natural sense of deterministic causation as a universal explanatory principle and with the deep and comforting belief, at the heart of many a philosophy or religion, that 'nothing happens for no reason'. The latter is tacitly prevalent even in science ('God does not play dice').

The relative absence of the notion of randomness in classical cell and molecular biology can be attributed to the fact that in the process in which events at the microscopic level generate a macroscopic behaviour, individual events that embody the stochastic process are 'averaged out'. The large number of cells used in most biochemical analyses, from Western blotting to DNA microarrays, eliminates awareness of cell-to-cell variability. Measuring in a cell lysate (collected from millions of cells) the concentration of a set of proteins will by default deliver the mean expression level of these cells and mapped into a deterministic scheme of explanation ($A{\rightarrow}B$). Such averaging is fine in many instances, such as when we measure the concentration of a hormone in the blood – the summing up of the behaviour of millions of cells that produced that hormone may even afford some statistical robustness. But the number of molecules of the same hormone in the cells of the gland that produces it can vary over a range of 10- to 1,000-fold between cell of the same nominal cell type, making it difficult to delineate the actual 'device physics' in a cell that mediate the hormone response.

Thus the law of large numbers that smooths out the 'erratic' processes of molecular collisions may come to help in some instances, but averaging fails in two situations: (1) where the behaviour of an individual cell, such as stem cell that can act as the *founder* cell of an entire lineage if not organ or a mutated tumour cell that gives rise to a tumour.[11]; (2) where the random fluctuations in cells can survive the averaging effect of the population[12] and even be amplified as the result of *interactions* between the members of the population that fluctuate, the molecules and the cells. This *channeling of noise* to control higher-level processes require non-linear interactions (as discussed below) and will result in the counterintuitive to randomness of macroscopic features, such as the random choice of cell types in development. The distinct fur patterns or fingerprint pattern between genetically identical individuals is the most prosaic manifestation of this phenomenon.[13]

Stochasticity is hard to accept and contradicts our sense of causality – even those of scientists – because our brain is wired for deterministic (and linear) causation, perhaps evolved to deal with macroscopic events in the environment. Eschewing randomnesshas profound cognitive roots also because it suggests not only the absence of law and order but also that there is no further 'upstream' causation. Randomness is where all the downward arrows of reductionist explanation have to end: things happen for no apparent reason. Then, by necessity, randomness would be the irreducible starting point of all explanations, the source of all driving forces.[14]

Non-Linearity in Gene Regulatory Networks

The second class of phenomena that accounts for the departure from linear causality is the behavior of 'non-linear' systems. The verbatim obviousness of the term 'non-linear' as the reason for defiance of 'linear causality' is unfortunate because the term has a specific meaning that derives from the study of non-linear dynamics and thus has independent roots from just being the opposite of 'linear' used so far to describe 'everyday causality'. Here we explain a set of elementary principles of non-linear dynamical systems, as embodied by gene regulatory circuits.

The Historical Case of the Toggle Switch

The first realization that the gene and its expression as a protein may not provide a straight ('linear') explanatory principle for the phenotype came early, right after the realization that genes regulate each other because they can encode regulatory proteins, or transcription factors. In the early 1960s, when gene regulation was discovered, Monod and Jacob immediately proposed that small circuits of genes that regulate each other could explain *differentiation* states of the cell,[15] as defined by a stable configuration of the expression status of the genes. In their example, two genes, *A* and *B*, repress each other via their encoded gene products, which are regulatory factors (see Figure 3.3). It is intuitively plausible that this mini-gene network can reside in two states. By 'state', a concept that will be central tool for the cognitive departure from linear causality, we mean the list of the gene activity values x_i for all the genes in the network, where here $i=A$ or B. Thus, the state of the network S can be written as the vector $S = [x_A, x_B]$. The two expected steady states are S_1, in which expression of *A* is high and *B* is low: $x_A >> x_B$ and S_2 with the opposite configuration of gene expression: $x_A << x_B$ (Figure 3.3).

Going into more detail, one can qualitatively see that the state S_3 in which both genes are balanced ($x_A = x_B$), is unstable: a slight increase in x_A, e.g. due to afore-discussed gene expression fluctuations, that temporarily affords a slight dominance to gene *A* ($x_A > x_B$) will increase the suppression that gene *A* exerts on gene *B*, further diminishing x_B relative to x_A. The inequality increases and expression of gene *A* further rises while that of gene *B* further diminish. This 'rich-get-richer' dynamics thus ensures that the state of equality $x_A = x_B$, is unstable.

But in this gene regulatory network the 'rich-get-richer' dynamics comes to a standstill at either S_1 or S_2 because of a feature that is generally (and reasonably) assumed in such circuit models: the degradation of the transcription factor *A* and *B* is assumed to follow *first-order* kinetics (explained in more detail in the later section 'More Detailed Study on the One-Dimensional Case'): its inactivation rate (dx_A/dt) is proportional to the concentration of the factor itself, x_A. Thus, once the rate of inactivation of A increases with the abundance of A, it eventually catches up since the increase in dominance of A over B will eventually slow down as B decreases towards zero ('diminishing returns'). While the increasing inequality follows the 'rich-get-richer' dynamics, the first-order self-degradation sets a limit – such that the entire system ends in either one of the two states, '*attractors*' S_1 or S_2 with the unbalanced expression levels, x_A and x_B (Figure 3.3, centre). These two states 'attract' the system (unless it is in a state that exactly satisfies $x_A=x_B$ and does not suffer noisy fluctuation, as discussed later). Each circuit can only be in one given state at a given time – hence if we were to start at any of state S_0 (except for when $x_A=x_B$), the system will be attracted either to S^*_1 or to S^*_2. The attracting property constitutes a convergence of initial states. All sets of initial states S_0 in which x_A is higher than x_B (by however minuscule an amount little) will be converge to S_1 and these states represent the '*basin of attraction*' of attractor S_1.

Implications for the departure from linear causality: If we apply causality to the states S of the network, we readily see the collapse of linear accusation: regarding the act of placing the system in a state S_x as the cause, then any of the initial state S_x in the basin of attraction of attractor S_1 will cause the system to assume the attractor state S_1 – this convergence of a set of diverse causes towards the same effect constitute the case (*II*) of the departure from linear causality. Of fundamental importance is the difference of the meaning of the arrows: in the causation scheme in [gene $A{\to}$gene B] on the one hand, and in movement of states [S_0 (initial state)${\to}S_1$ (attractor)] on the other hand. This will become clear in the next section.

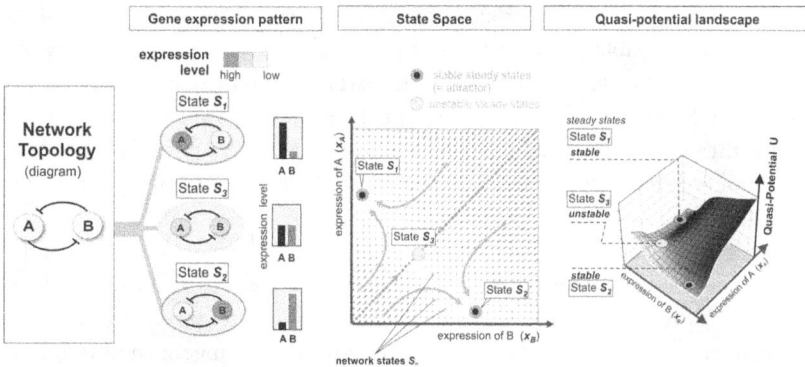

Figure 3.3: Two-gene mutual inhibition gene network and its dynamics ('toggle switch'). Gene network topology (left) for the two genes, A and B, that suppress each other and three example gene expression patterns, representing the three steady states (S1, S2, S3). Panels to the right of it: cell states as defined by the gene expression patterns. State space spanned by the two-gene expression variables xA and xB with vector field (small

**arrows), four trajectories (large arrows) visualize the overall flow towards the attrac-
tor states (dots with a dark centre); unstable steady steady-state (light-coloured dot)
on the boundary (diagonal) that separates the two basins of attraction. Right-most
panel: quasi-potential landscape (semi-schematic) where the elevation represents the
quasi-potential, such that the apparent 'energy barrier' represent the effort needed to
transition from one attractor to another.**

The State Space Formalism for Dynamical Systems

The entire dynamics described above can be framed in a mathematical formalism
which is described, still in a qualitative manner, in Figure 3.3.[16] The elementary
and crucial concept is that of the *state space* – the space in which every point
represent a state S of the system. The central idea then is that a system changing
its *state S* is equivalent to moving in *state space* (e.g. from an initial state to an
attractors as it executes a causation) to represent a process of a state change (Figure 3.4). And each motion has its cause – thus the driving force of the motion of
S in state space embodies a causation for the state change of a system. A fundamental distinction one needs to be aware of is that between the 'path' of the state
S and the 'pathway' in circuit diagrams that biologists are nowadays so fond of
drawing to 'explain' a biological mechanism. Such pathway diagrams consist of
nodes representing the genes/proteins (A, B) and their connections in the form
of arrows that indicate how one gene influences another gene. These influence
arrows are also understood as the arrows in the schemes of causation: increase
of A causes B to decrease. These pathway arrows must be distinguished from
the arrows that represent the path of a state S in state space– these two types of
arrows belong to distinct categories of symbolization!

The *state space* is space of all possible states, or – in our example – of all
theoretically possible configurations of [x_A , x_B]. This space is spanned by the
variables x_A and x_B which act as the coordinates of the state space to define positions of S. Since x_A and x_B also represent the expression values of the genes A
and B they also define a state of the system – hence the equivalency between
state and position in state space. The very concept of a state S embodies the simplest form of an 'emergent' property: its motion along *trajectories* in state space
is emergent because it results from the coordinated change of behaviours of the
two genes, the expression as captured by the variables x_A and x_B . The coordination is achieved by the network of interactions: they prevent x_A and x_B from
changing independently, thus forcing them instead to change such as to follow
predestined trajectories in the (x_A , x_B)-state space. A change in gene expression
pattern of a cell thus maps into a displacement of the state S from one position to
another. The concept of trajectories in state space visualizes the abstract notion
of coordinated change gene activities.

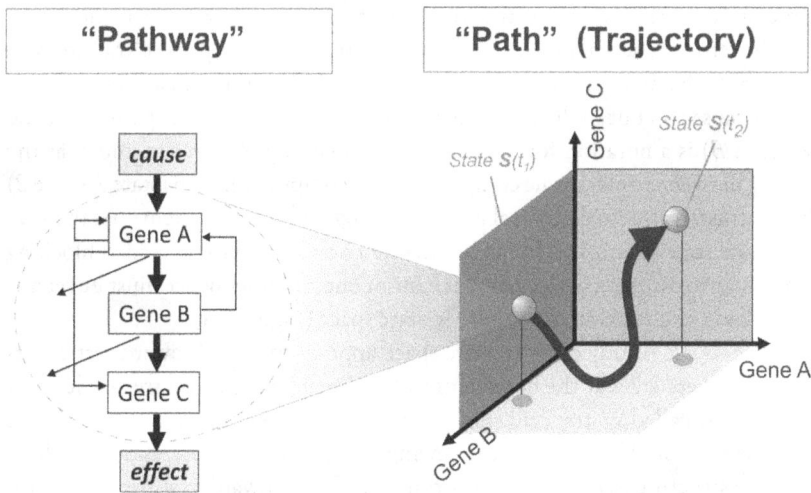

Figure 3.4: Conceptual distinction between a pathway in a signalling pathway diagram and a path in state space. The pathway (thick arrows top-down) in the pathway diagram ('arrow = arrow scheme') is often equated with the cause–effect schemes. By contrast the path = trajectory (curved thick arrow) in the state space represents a real physical process that drives the change of the system. Note the entire state of the pathway of network represented by the diagram on the left mas into one point in state space on the right.

Note that in network dynamics what changes is the configuration of x_A, and x_B values, that is, the state $S = [x_A, , x_B]$ whereas the network as an object remains unchanged in its structure (see Figure 3.3, left vs centre panel). In other words, the dynamics of S is governed by the specification of the interactions of the network (the wiring diagram or network topology) and the modality of interaction (e.g. whether an arrow means suppression or activation). Both are part of the network structure that embodies the system specification. Mathematically, the system specification, which prescribes the dynamicsis captured by a set of ordinary differential equations (ODE) (see below) which in this case is a two-dimensional system and describes how the rates of change for x_A and x_B, that is, dx_a/dt and dx_b/dt, depend on the values of x_A and x_B at a given time.[17] The nature of this dependence is described by the 'right-hand-side' terms of the ODE (e.g. see equation 1).

The notion of state space and trajectory now provide the conceptual framework to see the important fundamental difference between the molecular biologist's molecular 'pathway' and the developmental biologist's developmental 'path', as illustrated schematically for a 3-gene system in Figure 3.4. A pathway, such as $A{\rightarrow}B$, where A and B are genes that form a chain of regulatory interactions, unfortunately is all too readily mapped into a chain causation and mistaken as a process or a path. It is merely the structure of the system (e.g. the gene network) that governs the actual process which is represented by the path or trajectory in state space that in turn is driven by the network. Both path and pathway can represent causation but in distinct domains of conceptualization. A pathway (gene $A \rightarrow$ gene B) is a notation for a feature in the anatomy of a system, much as the wiring between to electronic components in a computer. A path (state $1 \rightarrow$ state 2) by contrast, is an actual physical process, through which the system goes to move from one state to another. In medicine, often a drug can be depicted as blocking a *pathway* to display a molecular mechanism but its consequence must be seen as a blockade or alteration of the *path* in state space (Figure 3.4).

The course of trajectory in state space appears to 'drain' towards attractors as if they were lakes at the bottom of valleys (see Figure 3.3, centre and left). In fact, the so-called vector field that indicates for any point acting as initial state S_0 where it would have to move following the system dynamics, appears to have a global structure that resembles coordinated 'flows' towards the attractors. This has stimulated mathematical efforts to depict the entire dynamics in the state space as a landscape, or more precisely, 'quasi-potential landscape' (Figure 3.3, right). Such landscapes were first proposed in the 1940s by Conrad Waddington as a handy metaphor to capture cellular development, but they in fact have a mathematical basis. However, this landscape is not an energy landscape in the familiar sense that the 'elevation' represents a potential energy that allows for a global comparison of energies and deriving the directionality of change. The reason is that a gene regulatory network, like all systems in living organisms, is not a classical conserved, 'gradient' or path-independent system. Hence we do not have 'true' potentials, and hence the use of the prefix 'quasi'.[18]

Attractor States and Multistability

The above two-gene mutual inhibition system is widely known as a 'toggle switch'[19] and has become a textbook example for non-linear dynamics. The central functional characteristic of the system is that the very same system can produce two attractor states – which Monod and Jacob likened to two 'differentiation states' because it offers a fundamental mechanism for how a system can differentiate itself into two distinct stable phenotypic states without mutations. This behaviour, first proposed for biochemical mutual-inhibition networks by Delbruck[20] even before non-linear dynamics became a discipline in its own right, is referred to as **bistability**, or more generally, **multi-stability:**[21] the very same system, defined by one single 'wiring diagram' of the network, can produce two or more attractor states, each defining a gene expression configuration. Thus, one single molecular anatomy produces multiple phenotypic states. The states are discretely distinct and intermediate states that, with respect to the configuration gene expression values, are instable. Grey is not stable, only black or white. Specifically, in our example we had the two stable attractor states, S_1, exhibiting the $x_A >> x_B$ configuration, and S_2, with $x_A << x_B$. These two states represent integrated phenotypes of the system (Figure 3.3). Note that each attractor is not only a steady state but also stable, in that random fluctuations in x which displaces S away from the attractor state S^*, if not too large, will be attracted back to the S^* – a manifestation that intermediates are not stable.

Multi-stability is deeply at the heart of the capacity of one genome to generate, via the gene regulatory network, a diversity of discretely distinct cell types, each defined by a type-specific gene expression profile S. For large gene networks of N genes, with N in the thousands, the same principles as described here, apply, such that each state is now a high-dimensional vector in the N-dimensional state space: $S=[x_1, x_2,...x_N]$ and possibly, thousands of attractor states are generated. In the 1970s Kauffman pioneered the studies of the generic properties of such networks using Boolean functions to capture the interactions between genes. He studied large ensembles of thousands of randomly wired networks from which generic features could be derived. Kauffman found that for relatively sparsely connected network and some readily satisfied structural design principles a large fraction of the ensemble exhibited multiple attractor states. He proposed that cell types in metazoan are attractor states of complex gene networks[22] – an extension of the basic idea that Delbruck and Monod and Jacob had suggested for the two-component circuits: the gene expression pattern across all the genes in a genome that defines a cell type would be encoded by a self-stabilizing high-dimensional 'state vector'. Such genome-wide profiles of gene expression can now be readily measured at the level of transcripts using new genomic technologies and is referred to as 'transcriptomes'.

Implications for the departure from linear causality: Each system that contains more than one attractor has a primitive memory system: an environmental input imposes/resets the initial state S_0, thereby placing the system state somewhere in state space into a given basin of attraction. Since a given environmental condition or associated signal influences the activities of a specific set of gene in a particular way, according to the signals' molecular nature, it is associated with a particular basin of attraction. Thus, environmental signals cause the system to move to a particular attractor S_1, S_2, or S_3, etc. In other words, the network with its characteristic quasi-potential landscape of attractor states classifies the environment. It remembers the class membership of external signals even after the signal has faded away because once the system has moved to an attractor it will stay there in the absence of other large perturbations. Put in simple, general terms: a transient cause A can have a lasting effect B that endures after the cause has vanished. Such persistence epitomizes the type (*IV*) departure from linear causality (Figure 3.1).

Unfortunately a great deal of molecular biologists to this date still cannot fathom this class of causal behaviour that obviously requires non-linearity. Therefore, they need to invoke a proximate cause that follows a linear causation scheme to satisfy their linear thinking and they need a cause that is embodied by a persistant molecular change, as conveniently epitomized by a genetic mutation – hence the dominance of genetic explanations in the mainstream molecular biology literature. Where, as increasingly observed, mutations cannot be found in gene sequencing, they typically resort to 'epimutations', the addition of molecular marks, such as methylation, on DNA or chromatin that does not affect the gene sequence. However, DNA methylation and other epigenetic modifications are nearly as dynamic and reversible[23] as any other molecular change, such as phosphorylation, because biological stability is determined by enzymatic regulation (which is subject to regulation by the network) and not by thermodynamic stability.

The Essence of Non-Linearity for Multi-Effect Causality

We have in the previous sections used the term '*non-linear*' without a definition. The term 'non-linear' is often loosely used to refer to some counter-intuitive behaviour not readily accounted for by a simple causal relationship; this is based on the general mathematical notion that a relationship between a possible cause A (whose quantity is represented by the horizontal axis of a graph) and effect B (vertical axis) cannot be graphed as a line, but rather a curve. But in conjunction with dynamical systems, 'non-linear' has a specific meaning. For instance, an exponential growth (such as microbial growth) which is evidently a non-linear behaviour in the mundane sense, is the result of a *linear* ordinary differential equation (ODE): the exponential function $x(t) = x_0 \, e^{kt}$ (where $x(t)$ is the number of microbes at time t) is the solution of the equation for the rate of change of x, namely, dx/dt in which this rate depends on x itself in a proportional manner: $dx/dt = kx$, where k is the proportionality constant representing the proliferation characteristic. Hence, exponential growth is the behaviour of a linear system. A non-linear ODE is an ODE in which the 'right-hand-side' that describes how the rate of change of the state variables x_i depend on each other, contains arithmetic products of the variables: they are multiplied by each other or themselves, e.g. the $dx/dt = ...$ expression that contains on the right-hand-side terms like: $x_1 \cdot x_2, x_1^2$ or x_1^3.

Multi-stability requires at least one positive feedback loop in the system structure and non-linearity in the ODE. (Note, however, that non-linearity, while necessary, is not sufficient for multi-stability.) In our toggle-switch example, since A suppresses B and B suppresses A we have the double inhibition which is equivalent to a self-activation loop. Multi-stability can be viewed as the emergent system behaviour that is promoted by non-linearity. Such behaviour can often not simply be 'read off' the network diagram interpreted in the mindset in which arrows represent linear causal or 'upstream/downstream' relationships as is still common in biology.

More Detailed Study on the One-Dimensional Case

In preparation for the discussion of one of the important but counter-intuitive principles of bifurcation, let us now simplify the system to a one-gene system with an explicit positive feedback loop (Figure 3.5): a gene A encodes a protein that activates its own expression and is degraded following first-order kinetics, that is, at a rate linear to its abundance, as previously explained. And again, this positive feedback loop will not lead to the exponential, uncontrolled explosion of cellular abundance x of the protein A: instead, x increases until it gets 'trapped' in the attractor state x^*_1 because of the degradation of A. It reduces x at the rate $(dx/dt)_{deg}$ that is proportional to x itself: $(dx/dt)_{deg} = -kx$ (where k is the degradation constant, the negative sign indicates negative rate, hence degradation). To make it interesting, for the synthesis rate $(dx/dt)_{syn}$ we do not use the first-order (exponential) function as used above for bacteria $(dx_A/dt)_{syn} = kx$, but instead a sigmoidal function. This is plausible and a widely made assumption because the self-activation is slow in taking off and will plateau for high values of x : $(dx_A/dt)_{syn} = x^4/(x^4+Q)$ where Q is the threshold or inflexion point of the sigmoidal function that relates the synthesis rate of x to x itself.

The synthesis and degradation rates jointly determine the dynamics through the following one-dimensional ODE:

$$\frac{dx}{dt} = \left(\frac{dx}{dt}\right)_{syn} - \left(\frac{dx}{dt}\right)_{deg} = \frac{x^4}{x^4+Q} - kx \qquad (1)$$

such that the net rate of change of x, dx/dt, which is the result of synthesis rate (syn) minus degradation rate (deg), exhibits an interesting change of signs along the axis x from which the net change dx/dt depends. The key task in analysing this model is to examine how the *rate of change of the state variable, dx/dt,* depends on the system state (here $=x$) itself. The graph in which the system state x determines the rate of its own change has a central meaning, we call it the *characteristic curve* of the system: it is the dynamics interpretation of the 'blue-print' or anatomy (Figure 3.5, bottom) of the system that is represented by the circuit diagram and the ODE that describes it and that fully determines the behavioral repertoire of the system. We obtain the *characteristic curve* by plotting $dx/dt = F(x)$ according to the right-hand-side of equation 1 as function of the state variable x. Where it crosses the $dx/dt =$ zero line, we have a rate of change of zero $=$ or no change indicating a steady-state. If the curve $F(x)$ crosses the zero-line multiple times (that is, $F(x) = 0$ has multiple solutions, we have multiple steady-states: Mathematically in our case this is achieved by a non-linear (in this case the steep sigmoidal) dependence of the synthesis rate of protein A $(dx/dt)_{syn}$, from its own abundance value, x, jointly with the linear increase of degradation $(dx/dt)_{deg}$ with increasing x. This relationship is graphically shown in Figure 3.5. The figure shows that for the net behaviour we can find three points in the state space x_A for which $dx_A/dt = (dx_A/dt)_{syn} - (dx_A/dt)_{deg} = 0$. Thus, the system has three steady-states: S_1, S_2 and S_3 at the values x^*_1, x^*_2 and x^*_3. A more detailed analysis (so called, linear stability analysis, see also Figure 3.5), shows that only x^*_1 and x^*_2 are stable, that is, attractor states. By contrast, while S_3, with $x=x^*_3$ is an unstable steady-state: a slight departure from it, e.g. due to gene expression noise, will push the system towards S_1 or S_2. This introduces the central idea of *instability* as a state that is sensitive to perturbations.

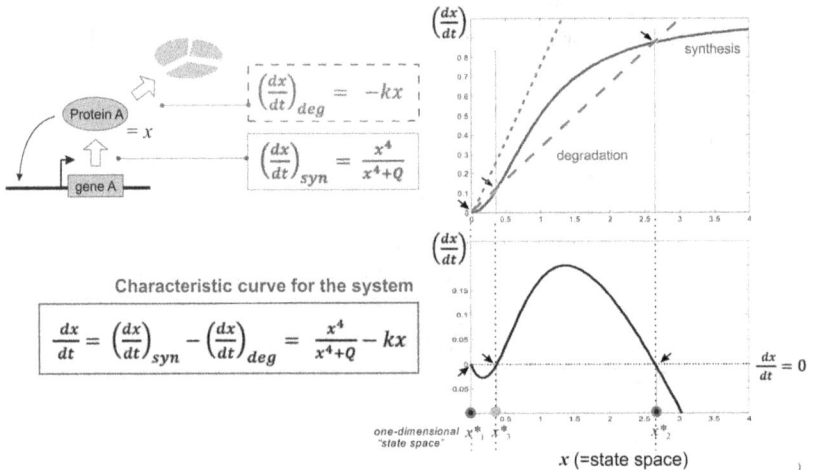

Figure 3.5: Graphical analysis of the ODE for the one-gene auto-regulatory system. Schematic representation of the system structure ('network') consisting of two steps: self-activating synthesis and degradation of protein A, its abundance, x is the state variable. The ODE for this system has two components, synthesis and degradation. The rates of these two processes as function of x itself are depicted (as absolute values) on the upper graph. The lower graph is the entire function (right-hand-side of ODE) and represents the characteristic curve for the system. Intersect of the degradation and synthesis functions – equivalently, in the bottom graph for the entire function, the roots (intersect with $dx/dt=$ 0 line) indicates steady-states. x^*1, x^*3 (circles) represent the stable steady states (attractors); x^*2 is the unstable steady state. Whether a steady-state is stable or unstable can be determined graphically: if departure from the steady state in the positive direction (higher x values = to the right) encounters negative values for dx/dt (= vertical axis) and departure to lower x-values encounters positive dx/dt values, then that steady state is stable, etc.

Bifurcations

With the above conceptual tool we can now describe bifurcations, a fundamental principle of non-linear dynamics which is even further beyond the reach of the intuition of linear causation than the above concepts. We do so using again the simple one-dimensional system and without use of mathematical equations. A bifurcation is a behaviour that appears when one gradually changes a parameter on the right-hand-side of the ODE, in our case, the constants k or Q, while monitoring the associated change of the steady states of the system.

Let us alter the degradation rate constant k: if the decay rate is low, auto-stimulation of the protein A will for a given value x at time 0, $x(t_0) = x_0$ (the initial state of the system) drive the self-propelling movement of the system towards higher and higher levels of A, but will meet the limitation by the self-degradation at some point: the net change $dx/dt = (dx/dt)_{syn} - (dx/dt)_{deg}$ is initially positive and high but decreases to zero because: $|- (dx/dt)_{deg}|$ increases faster than the synthesis rate $= (dx/dt)_{syn}$ as x increases (since the latter suffers from saturation – as manifest in the sigmoidal curve). The system enters the attractor state S_2 when the degradation rate meets the synthesis rate: dashed line $-kx$ crossing solid curve $+x^4/(x^4+Q)$ in Figure 3.5, top). If we lower the degradation rate constant k by bit a (making the line in Figure 3.5 flatter) and rerun the system again from the same initial point $x(t_0)$ it is geometrically clear from the graph (Figure 3.5, top) that the intersect that determines the position of attractor S_2 will be shifted to higher values of x: it requires a larger amount of x for the now-weaker degradation to neutralize the auto-stimulation. As a consequence, the attractor state S_2 is defined by a higher vale x^*_2.

In the opposite extreme case, the self-degradation constant k can be so high that A can never take off, independent of its initial position: a slight temporary dominance of synthesis will be in the 'flat region' of the sigmoidal curve of $(dx/dt)_{syn} = x^4/(x^4+Q)$ and will be met by a massive increase of degradation of A. In the same graph (Figure 3.5, top) the dashed degradation line is so steep that it is always, that is, for all values of x_0, higher than the solid curve for the synthesis. Thus, if k is very high the system has only one steady state: $x^*_1 = 0$ – obviously no protein is ever present. This steady-state is stable: any temporary increase of x (due to a chance fluctuation) will now be countered by the system since degradation immediately sets in and is always higher than auto-stimulation: the system state is attracted back to the $x^*_1 = 0$ state.

In conclusion, the system can exhibit qualitatively distinct system behaviours depending on the value of the parameters in the ODE, here the degradation constant k. It is important to emphasize the following two hallmarks of our analysis: first, k is a constant, as opposed to the variable x which changes in 'real time' of the observed system behaviour – yet we artificially altered it. For each value of k, we evaluated the system dynamics – as if for each different k we had a different system. Thus, k represents a system characteristic and influences the above discussed *characteristic curve* of the system: depending on its value, the system can have only one stable steady state, at $x^* = 0$, or, for lower values of k as, two attractors, as discussed initially. Second, for each value of k, we 'run' the system by allowing the ODE to operate on the position of the initial state $S_0 = x_0$ and let the system evolve until it reaches an attractor state. Thus change of the system parameter k and of the state variable x occur at different time scales: the variables change in real-time, for a given constant value of parameters. The parameter k changes at a much longer imaginary time scale, as if we switch to a different universe in which the value of k is different. This *separation of time scale* is traditionally assumed a priory and the constant parameters, such as k, Q, in the ODE are referred to as *control parameters*. For gene circuits, each set of constant parameter values is part of the system characteristics and is encoded in the genome, such as the inherent stability (degradation of a protein at a given rate). Therefore, the much longer time scale in which the control parameters, hence the system characteristics, are altered is equivalent to *evolutionary time scale*. Change of parameters are most readily imagined as genetic mutations which, for instance, may impact the structure of protein A such that its degradation is affected or it regulates its target gene differently. (Here we do not further venture into causation in the realm of evolutionary biology which is a different matter.)

With the notion of change of control parameters and observing its impact on system behaviour, we can now draw the *bifurcation diagram* (Figure 3.6): The dependence of the *position* of steady states, that is the x-values that satisfy the condition $dx/dt = (dx/dt)_{syn} - (dx/dt)_{deg} = 0$, as well as their *stability* (stable vs instable steady state) as function the value of a control parameter. In our one-gene system, we focus on the control parameter k and theoretically imagine a continuous sweep of k from 0 to higher values, and plot the positions of stable and unstable steady states for each value of k (see Figure 3.6, left). At certain values of the control parameter, an attractor state may disappear, it may be converted into an unstable steady state, or jump to another position. Such sudden qualitative changes of the behavioral repertoire of a system are called *bifurcations*. They separate various behavioral regimes in the *parameter space*.

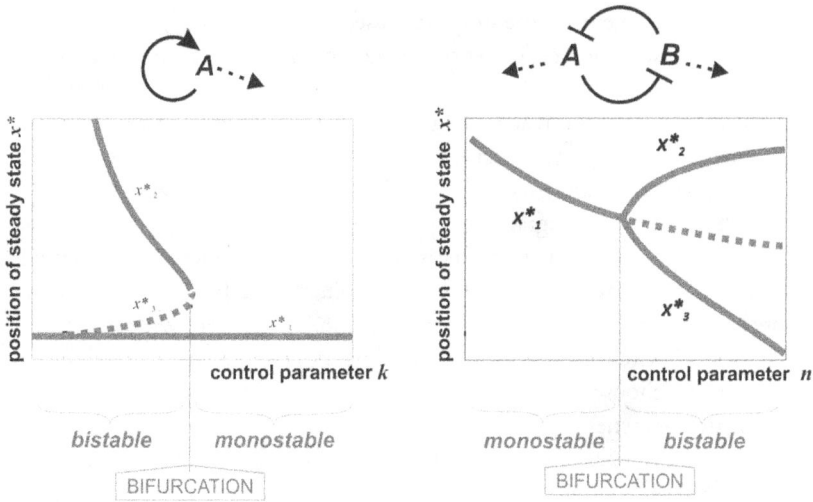

Figure 3.6: Bifurcation diagrams for the two systems discussed. Right panel: bifurcation diagram for the one-gene circuit of Figure 3.5; left panel: bifurcation diagram 'pitchfork bifurcation' for the toggle-switch discussed in Figure 3.3. Thick lines indicate the state space position (vertical axis) of the stable steady state as the control parameter *k*, or *n* (horizontal axis), respectively is altered.

Bifurcation diagrams can be drawn from higher than one-dimensional systems and can take all kinds of shapes. For our two-gene system (Figure 3.6, right) the positions of the steady states with respect to just the gene *A* dimension, that is x^*_A, are shown. Here we can see that as the Hill-coefficient *n* (which characterize the steepness of sigmoidal function that characterizes the mutual inhibition) increases, the system behaviour changes from a monostable behaviour to a bistable behavior at the bifurcation point which takes the shape of a branching into two 'stable branches' as it becomes unstable, lending such generic qualitative jumps of system behaviour the name 'bifurcation'. Even more figuratively, the bifurcation of the kind shown in Figure 3.6 (right) is referred to as a *'pitchfork bifurcation'*. But more generally, we see here a fundamental feature of bifurcations: they permit us to explain how a gradual quantitative change of a parameter that characterizes the internal wiring of system, can sometimes result in a sudden qualitative change of an emergent property of that system, such as system behaviour.

Causality Due to Bifurcation Behaviour

The effect of a change of parameters on potential behaviours of a system opens new dimensions to causality. So far we have discussed how change in one variable x_A (gene *A* expression) influences another variable x_B (representing expression of its physical target, gene B), the simplest cause–effect pair, for a linear upstream/ downstream system. At the longer time scale we can use the change of a parameter k as a cause, with the effect that its change may alter the boundaries of the behavioral regime. This is readily imagined in the characteristic curve of the system: change of its shape will change the position and existence of intersects with $(dx_A/dt)_{syn} = 0$, and can also convert a stable into an unstable steady state. Note in that the bifurcation diagram each position in the parameter space represents a distinct characteristic curve of the system, hence a distinct system anatomy with a distinct behavioral repertoire. In the pitchfork bifurcation crossing the bifurcation point (from left to right in Figure 3.6, right panel) we destabilize the attractor x^*_1 (switch from solid to dashed line in the bifurcation graph) forcing the system to 'choose' between the two now newly appearing attractors x^*_2 or x^*_3 –the two branches.

Implications for the departure from linear causality. The pitchfork bifurcation example (Figure 3.6, right graph) illustrates the case (*III*) departure from linear causality: if the cause A is the change in parameter space, it puts the system (in attractor state x^*_1) in a regime that is susceptible to additional causative factors that determine between the two outcomes, B and C (represented by the two new attractors x^*_2 or x^*_3) which were not even in the realm of the accessible before the cause A has 'acted'. Whether the fate B or C is chosen depends on whether the additional signal which increases or decreases x (and which can be deterministic or stochastic). Note that change of A occurs at the slow time scale of a parameter change.

The case of stochastic system is of interest here: x randomly fluctuates (but was held in the attractor of the monostable regime). The cause A shifts the behaviour into a regime in which the two well-defined, distinct binary outcomes B or C become randomized – this is an example of the propagation of microscopic stochasticity (random fluctuations in x) to higher 'emergent properties' without averaging out – thanks to the determinist constraints embodied by the interactions with the system and manifest in the bifurcation. Because gene expression noise enforces an irreversible decision to B or C, the decision is intrinsic, driving a 'diversification from within'. It 'breaks the symmetry' of the original attractor state, which was 'agnostic' and hence symmetric to the possibilities B and C, by forcing a binary decision between the two options. Such symmetry-breaking behaviour is fundamental in the development of complex organisms which has been viewed as a sequence of symmetry-breaking bifurcations.[24]

The Old Dualism of Selection vs Instruction

The pitchfork bifurcation helps illustrate the interesting distinction between instruction and selection – the two alternative modes of causing changes in evolution as well as in cell lineage determination during development.[25] While this subject goes beyond the scope of this chapter, a concrete example from stem cell biology should illustrate the concept. In the fate decisions of multi-potent progenitor cell, a common progenitor cell $\alpha\beta$ (e.g. of the blood) can differentiate into the α or the β lineage (e.g. the red vs the main white lineage) as dictated by the gene regulatory network. A specific change in the progenitors' biochemical environment[26] can result in the destabilization the progenitor state without specifying the lineage – in which case the cell becomes susceptible to the cell-external fate-determining factors a and b which impart a deterministic push in either direction, to fate α or β, respectively. For instance, the soluble cytokine EPO promotes the red cell (erythroid) lineage and the cytokine GMCSF stimulates the white cell (myeloid) linage in the progenitor cell, notably when its state is destabilized.[27] This scenario in which a parameter change establishes the conditional susceptibility to causal factors (a, b) illustrates the complications that one faces when reducing processes that obey a non-linear multistable dynamics with a structured parameter space (bifurcation diagram) to a linear causation captured by an arrow–arrow scheme.

There is a twist on top of this regulatory principle: The fate-determining factors a and b, as developmental and cell biologists have long noted, often not only influence the fate decision of the (destabilized) $\alpha\beta$ progenitor cells (hence, 'cause' a fate), they also act as survival and growth factors of the respective committed cells: a expands the few cells that *by chance* have entered the α fate (and as part of its new gene expression profile, expresses responsiveness to the factor a), and analogously the factor b expands cells just committed to the b fate.[28] This *selective* expansion is akin to the Darwinian principle of natural selection in which a randomly acquired variant phenotype experiences a preferential propagation in a giving 'selective' environment. Thus, in addition to deterministic instruction, where the concept of linear causation might in some cases be appropriate, cell fate determination is also established by selection of randomly acquired states. The long controversy as to whether cell fate determination is instructive or selective[29] is moot if one sees the behaviour through the lens of non-linear stochastic dynamics around a bifurcation event that opens access of a stochastic system to two new attractor states: both are complementary to each other and not mutually exclusive. However, many molecular biologists habituated to mapping linear causation onto signalling pathways, have difficulty embracing the act of selection in fate determination in which a random microscopic process plays a constructive role in the generation of a complex, ordered gene expression pattern. This was similar with accepting Darwin's natural selection of random variants in the early days. Perhaps understanding that the two alternative choices are robust attractors, predestined by the non-linear interactions, may help.

Concluding Remarks

Causal thinking in molecular biology is preoccupied with identifying the proximate, that is, the molecular cause because it offers a handle on how to influence phenotypes with drugs. We have dealt with the change of cell states and revealed the difficulties encountered when molecular pathways, the physical actuator of such change, are mapped into schemes of causation ('arrow–arrow' schemes) and vice versa. Such cognitive procedures are successful only in exceptional cases: on the islands of deterministic linearity within the sea of stochasticity and non-linearity. But the latter two do not simply blur any straightforward, plausible causal relationship. Because of interactions, there is no 'averaging out', as many systems form larger systems. Instead, interactions in systems of any interacting components, such as the genes and molecules that exhibit non-linearity, are constructive in developing 'higher level' features, the valleys on the epigenetic landscape, that guide entire cell populations to distinct phenotypic states. The cells of various but well-defined types in turn interact to form tissues and organs, which influence each other through a network of hormonal and neural influences. This multi-layer hierarchy is a network of networks, nested into each other to form a vertically integrated system that operates causal networks at multiple size scales from the microscopic to the macroscopic, and across size scales, each with its own dynamics and attractor landscape. The interaction between the entities represented by attractors of the various networks, cells tissues, organs, organisms, societies, results in multi-scale causality which poses another challenge to the linear, proximate causality that still dominate the thinking of life scientists and yet has to be parsed in the same way we have shown here just for the level of molecular networks.

4 EMBODIED INTELLIGENCE IN THE BIOMECHATRONIC DESIGN OF ROBOTS

Dino Accoto,[A] Cecilia Laschi[B] and Eugenio Guglielmelli[A1]

From Biomechatronics to Biorobotics

Robotics is a relatively young discipline, whose origins are set in 1960 when the first robot was installed in a General Electrics production plant in New Jersey. Robotics developed first and foremost in the manufacturing industry, but its potential for applications in service tasks was soon recognized.[2]

Robot design is based on the mechatronics design principles. The word 'mechatronics' merges *mechanics* with *electronics*, in the same way as modern machines integrate electronics, sensors and control. Mechatronic design is, then, the integrated design of different components. A typical mechatronic system is composed of a mechanical part (usually an articulated structure with many degrees of freedom, in case of a robot); of a number of proprio- and extero-ceptive sensors; of actuators; of energy sources; of a network of microprocessors (usually 'embedded') and of analogue and digital signal processing boards; of control interfaces; and of communication units.

Similarly, biomechatronics can be defined as the concurrent engineering approach combining, in a synergistic way, information and methods originated by control engineering, mechanical engineering and life sciences.

In case the same engineering problem, as it is set out by its technical specifications, admits more than one design solution, it is desirable to point out, among all the possible solutions, the *optimal* one. The optimization of heterogeneous systems is generally based on global multi-domain models, which must be sufficiently refined to quantitatively describe the functional links among the subsystems, highlighting the interactions that contribute to the achievement of the overall performance. It is quite frequent that the complexity of the model, for example because of the large number of parameters involved, possibly interrelated through functions that are not known or suffering from high levels of uncertainty, makes it difficult or even impracticable to identify a global optimum. In this case, it may be useful to resort to sensitivity analysis. If P_i $= f_i(q_1, q_2,...,q_n)$ is the *i-th* performance of the system, function of the n parameters q_k *(k=1 ... n)*, the sensitivity of p_i with regard to the generic parameter q_j is defined as: $s_{ij} = \frac{\partial f_i}{\partial q_j}$. Evidently, if S_{ij} is a positive number, the value of q_j should be increased, if possible, in order to improve p_i. Conversely, if $S_{ij} <0$, p_i will increase if q_j is reduced.

Pursuing the optimal design not only maximizes performance, but it brings, as a corollary, a further important benefit. Let's suppose that p_i is a regular function of the design parameters q_1, q_2... q_n, i.e.: $p_i = f_i(q_1, q_2..., q_n)$. If the design is *optimal* for a given set (Q) of design parameters, , then the sensitivity of p_i with regards to all the design parameters is zero in a neighbourhood of Q. The demonstration is trivial, since it is a direct consequence of the fact that $\lfloor grad\, f_i \rfloor =0$ in Q. This implies that small changes in the reference values of the parameters (for example, due to deformations, wear, manufacturing defects, fluctuations in environmental parameters, etc.) do not alter the considered performance in a significant way. Therefore, the optimal solution to a given problem is also the most stable with regards to small changes of the design parameters.[3]

It is evident that, in order to benefit from the stability of the optimal design, it is necessary that the model of the system encompasses all the parameters that contribute to the performance p_i. Only some of these parameters are true design variables, i.e. parameters to which the designer can assign a desired value (e.g. geometrical dimensions), while others refer to the environment in which the device operates. The environment is described in terms of quantities that can be: (1) physical constants (e.g. gravity, Boltzmann's constant, the speed of light); (2) design parameters that may be considered fixed for the specific application (e.g. friction between parts in relative motion; ambient temperature; anthropometric measures of a specific user's body) and (3) unknown variables outside the control of the designer (e.g. distribution of rocks in a soil, position of passers in a road, posture of a healthy subject). An environment in which important parameters can assume a value not known a priori is said to be*unstructured.*

There is a dichotomy between the subset of design parameters to which the designer can assign a value, and the subset of design parameters that the designer must consider as assigned (i.e. *external* design parameters) and, as such, outside of the *design domain*. In the case of robots intended to operate in unstructured environments, the designer must incorporate in the machine a sufficient level of autonomy to cope, through adaptation, with the changing environment. In fact, in robotics, 'intelligence' often stands for computational capabilities, necessary to allow the adaptation of robot behaviour to internal (extero-) and internal (proprio-) stimuli.

In the big move from industrial robotics to service robotics, a strong need emerged for robots to perceive their environment (which cannot be structured like in industrial scenarios), to interpret the sensory information and to plan their motor behaviour accordingly.[4] Looking at the way nature has developed organisms perfectly able to live in such environments appears straightforward.[5]

From this need for robots able to negotiate real-world environments and uncertainty, together with the need and the capability of applying robots in the biomedical field, robotics and life science met in so-called *biorobotics*. From a methodological point of view, biorobotics consists of the following phases (Figure 4.1):[6]

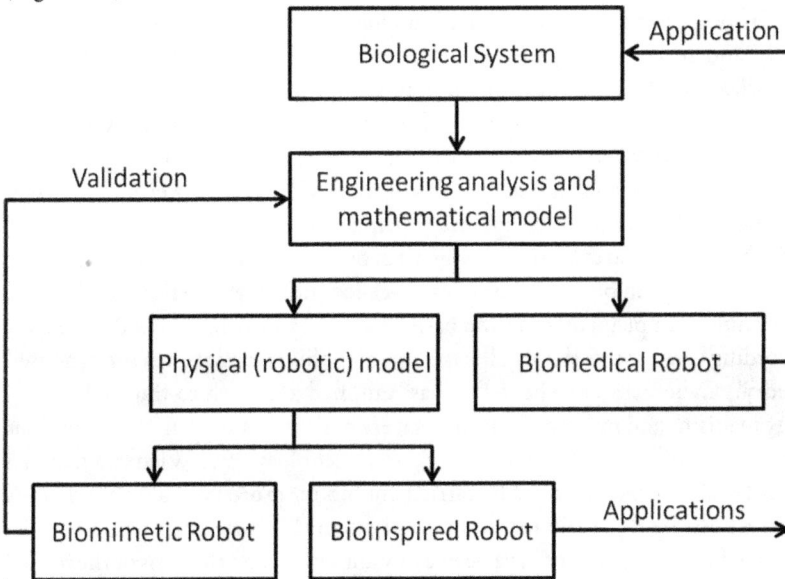

Figure 4.1: The biorobotics methodology.

The first phase is studying and modelling the biological system with the methods and tools of engineering and robotics. Such a biological system could, for example, be the human beings on which the robot under development will operate, as in the case of surgical or rehabilitation robots (biomedical robotics); in this case the next phase is to use the model developed in the first phase for designing the biomedical robot, and in particular for designing the interaction with the human being. The method then loops on the biological system itself. Alternatively, the biological system can be either the human model or a replicated animal model. The next phase is then designing the robot according to the model developed in the first phase. Such design can be used in two ways: for the development of bioinspired robots for diverse applications, and for the development of a physical model of the biological system (biomimetic robot) to use for further understanding of the biological system. The method then loops on the modelling itself.

Bioinspired and biomimetic robotics is the wide area of robots designed and built in imitation of living organisms, be they humans or animals, or other types of living beings. As explained, the development of these robots has a twofold objective: to solve problems of real-world negotiation and to validate models of the living organisms they are inspired by. In this broad area one finds *humanoid robotics*, where robotics technologies are advanced both for building human-like robot bodies and for giving them adequate perception and cognitive capabilities. Inspiration from animals is also bringing strong innovation in the materials, shape and mechanisms of robot bodies, as well as in the bodies'control and sensorization. A current line of development in this research area is *embodied intelligence* (see next paragraph), i.e. how much of the control is given by the shape of the body and the mechanical characteristics of materials.

Biorobotics science has deep roots in history. In fact, it is even older than robotics itself. The use of machines to explain the behaviour of living beings dates back at least to cybernetics, which can be considered as established by Wiener in 1943.[7] Even before cybernetics, so-called proto-cybernetics stated that 'if a machine is implemented on the basis of a theory of behavior, and it behaves according to what this theory allows to predict, this test reinforces the proposed theory'.[8] Cybernetics can be defined as 'a unified approach to the study of living organisms and machines'. Machines were built as 'material models' useful for testing scientific hypotheses. According to Rosenblueth and Wiener, a material model enables experiments to be carried out **'under more favorable conditions than would be available in the original system'**. It may enable results to be obtained which 'could not have been easily anticipated on the basis of the formal model alone'. A material model is not always biomorphic since 'it should imitate the natural function or behavior, not necessarily the appearance of the living organism featuring that behavior'. Machines were then used for science. This approach can be used for corroboration/falsification of theories by testing if the natural and the robotic systems behave in the same/different way under the same external and internal circumstances. It can also be used for deciding between two competing hypotheses: if two theoretical models, M1 and M2, are both implemented on robots and if the behaviour of the robot built according to the theoretical model M1 is more similar to the modelled biological behaviour than the behaviour generated by the robot built according to the theoretical model M2, then the M1 hypotheses is strengthened, and vice versa. Finally, biorobotics can be utilized for generating new hypotheses on the functional structure of the biological system.[9]

More recently, robotics technologies have provided more suitable tools for replicating living organisms and for demonstrating or validating models and hypotheses about their behaviour.[10] Actually, taking inspiration from Nature is older than robotics. The word 'biomimetics' was firstly used in the 1950s by Otto Schmit, to indicate 'the application of biological methods and systems found in nature to the study and design of engineering systems and modern technology'. Indeed, examples of bioinspiration appear in many fields of engineering different and older than robotics. For instance, truss structures, used since ancient times in buildings construction, combine high robustness and low weight and closely resemble the trabecular architecture of birds bones. Such structures allow us to solve several engineering problems, and some of them, such as the Eiffel Tower in Paris, still arouse admiration. In another example, the widely diffused Velcro© adhesive comes from the observation of how well burrs attach to animal fur.

However, taking a biomimetic approach in robotics does not mean blindly copying nature. The tool of nature is evolution: organisms can change over generations to adapt to their environment. Evolution guarantees that a solution is found for survival in a given environment, but it does not guarantee that it is the optimal solution. On the contrary, it is an incremental improvement which keeps traces of the past configurations. In the words of Robert Full, head of the Poly-PEDAL Laboratory at Berkeley University, California (USA), and one of the first scientists in biomimetic robotics,

> **We think blind copying is exactly what you don't want to do.** You will fail miserably, because nature is way too complex ... Rather than seeking to copy any specific morphological or even physiological detail, we hypothesize functional principles of biological design and test their validity in animal and physical models.

Pursuing bioinspiration in robotics means deeply analyzing the problem the robot has to respond to and then the solutions that Nature seems to have adopted to respond to the same problem. Understanding and replicating the working principles of such solution is the key method for biomimetic robotics.

Morphological Computation and Embodied Intelligence

As recently highlighted,[11] the embodied intelligence paradigm, by reevaluating the role of matter in linking perception and action, blurs the rationalist approach to artificial intelligence. In this perspective, matter, which assumes an active role in the generation of useful behaviours, can be considered *smart* but not *intelligent*, not in the cognitive or in the strictly etymological sense. This is because the link between perception and action that it manifests is devoid of any aspect of intentionality. From this point of view, the kind of intelligence contemplated in the embodied intelligence paradigm must be regarded as a form of adaptability, which is useful to negotiate real-world scenarios.

Adaptation stands for the capability to plan, through proper computation, the best way to act in a given state and situation. To go deeper in this overview, we now have to investigate what computation is and what performs computation. According to the definition given at the International Conference on Morphological Computation (ICMC2007), a process can be called a *computation* if: (1) we can systematically identify a relation between input and output; (2) the process is programmable for the generation of different classes of non-trivial input-output mappings; (3) the process is useful for some specific purposes. Computation is usually performed by a computer, a machine capable of executing a programme, which is the translation of an algorithm in a specific language. A computer basically manages logic expressions. Recent findings show that it is not only computers that can compute logic expressions –purely mechanical systems also can. More specifically, it has been demonstrated that computation of logical expressions, after an adequate codification, can be executed by simple machines through their specific structure (morphology) and their interaction with the environment.[12] According to Pfefer and Lida,[13] morphological computation consists in 'tasks distribution among the controller, the morphology of the agent (shape, sensors, actuators, materials), and the environment'. If morphology (i.e. hardware) is capable of performing (at least part of) computation, than programming is not just a matter of writing code but we may think of achieving useful behaviours also by properly designing the 'body' of the robot.

This perspective leads to the diaeretic conclusion that, if it is true that there are external design parameters, some of them can still be exploited by the designer for the furtherance of her/his purposes. The external parameters, related to the environment, affect the performance of the system through the physical interaction between the environment and the construct.

The design problem then becomes that of identifying effective techniques to design not only the biomechatronic system, but also the physical interaction that the system will establish with the environment in which it operates.

In this perspective, the emerging design paradigm of embodied intelligence aims at allocating some basic control function to the 'body' of the machine, thus significantly simplifying the higher levels of control and reducing the need for embedded sensors and electronic computational units. More in detail, the embodied intelligence paradigm aims at exploiting the benefits deriving from the allocation of part of the computation involved in the performance of a task to the mechanical structure of a robot. Such benefits include:(1) simplification of morphology and control; (2) achievement of fast response times; (3) increased adaptability to the environment.

While there is no doubt about the potential benefits of embodied intelligence for the design of machines capable of interacting with unknown working environments, a fundamental question arises concerning *the way* designers can embed intelligence in a structure. In fact, conventional engineering design methodologies, based on system models, are not suitable to exploit embodied intelligence, because morphological computation occurs in the interaction of the machine with the environment, where dynamics plays a central role, and designing a machine starting from its desired dynamical properties is not at all trivial from a modelling point of view.

The design methods that can be pursued to overcome traditional calculation-based optimal design problems vary, as one may expect, depending on the type of systems and on the type of environment they will operate in.

In the following, we will analyse two case studies, in which the environment is respectively in the physical world (case study 1) and the human body (case study 2).

The Octopus as a Model for Embodied Intelligence in Soft Robotics

The octopus is a marine invertebrate with amazing motor capabilities and intelligent behaviour, which is quite impressive considering its position on the evolutionary scale (Phylum: Mollusca, Class: Cephalopoda). Most recognized theories explain that these enhanced behaviour and capabilities for interaction with the environment are due to the special morphology of the octopus body, and especially due to the high manoeuvrability attributed to the form and materials of the arms and their efficient neural control mechanisms. The octopus represents a biological demonstration of how effective behaviour in the real world is tightly related to the morphology of the body. The octopus moves by crawling and swimming, has reaching and grasping capabilities, camouflage abilities, and smart strategies for hunting and for hiding.

Embodied Intelligence in Octopus Movements

Some of the octopus movements have been well characterized and modelled.

Reaching is the movement of the octopus arm used to reach a target and then grab it. It is based on a bending wave that travels from the arm base to the arm tip.[14] The bending is sharp enough to minimize the effect of water resistance on the arm movement and the drag force. Despite the complexity of the octopus arm muscular structure, the virtually infinite number of degrees of freedom, the possibility for the octopus to control many muscle bundles independently, the control of the reaching movement is controlled by a very small number of parameters, thanks to simplification mechanisms and stereotyped movements.

Fetching is the movement by which the octopus brings the grasped object to the mouth, which is at the centre of the eight arm bases.[15] In this case, too, a couple of bending waves travelling from the arm base and from the arm tip, respectively, simplifies control to a small number of parameters and optimizes the effect of the interaction with the water.

In *locomotion* over a substrate, the octopus puts in place a mechanism by which some of the arms are used as a sort of leg. Interestingly, the arms used are a couple of back arms, which push the body forward; thanks to the neutral buoyancy of the octopus, it allows the body to be kept raised from the terrain.[16] In *swimming*, the octopus uses a pulsed-jet propulsion obtained by contractions of the mantle, which expel water from a funnel. It has been demonstrated that the properties of the mantle and its interaction with water optimize the propulsion. The effect of water on the compliant mantle deforms the mantle to an extent depending on the speed. The new shape, reduced in wideness, reduces friction and improves speed.

Biomechatronic Design of an Octopus-Like Robot

According to the biorobotics methodology, the observation and modelling of the biological model is the first phase to accomplish, by using engineering tools and viewpoint. In the case of an octopus, the key principles for the arm dexterity and simplexity of control have been investigated, together with insights on the anatomy, neurophysiology and biomechanics of the octopus arms. The measurements of the arms of twenty-four octopus specimens are presented in detail[17] outlining the following findings:

- the octopus arm can elongate by 70 per cent with a 23 per cent diameter reduction
- the average pulling force of one arm is 40 N
- each arm can shorten of 20 per cent on average, at a rate of 17.1 mm/s
- the longitudinal stiffening rate reaches 2 N mm/s.
- the nerve cord has a sinusoidal arrangement inside the arm, compatible with arm elongation abilities
- the insertion points of the longitudinal and transverse muscle fibres confirm local contractions.

The observation of the locomotion strategies of the octopus brought us to understand that the back arms are used to push the body forward,[18] while the octopus pulsed-jet swimming has been studied and modelled with fluido-dynamics tools.[19]

The knowledge on the octopus arm led to the design of a robot arm with:

- longitudinal and transverse contraction units,
- an external braid transmitting the deformations from the local insertion points of actuators to the rest of the arm, thanks to crosslinked fibers forming same angles as connective tissue fibers in the octopus (68°–75°),
- electric wires arranged in a sinusoidal way inside the arm,
- transverse actuators (SMA springs) arranged radially in the arm diameter,
- longitudinal actuators (cables and SMA springs) of different lengths, to allow local contractions.

An Octopus-Like Robot Exploiting Embodied Intelligence

The resulting eight-arm soft robot shows some of the key principles of the octopus embodied intelligence, and specifically of the octopus behaviour, movements and control.

The octopus robot is a completely soft robot, which integrates eight arms extending in radial direction and a central body, in which the main processing units are contained (see Figure 4.2).[20] The front arms are mainly used for manipulation, elongation and grasping, while the others are mainly used for locomotion. To optimize elongation, reaching and manipulation tasks, the front arms are based on the SMA actuators, which reproduce the internal anatomical features of the real octopus arm, and thus allow them perform finely controlled and precise movements. The other arms, which are used for crawling, are made from silicone and cables, embedding the features needed to obtain an octopus-inspired locomotion. The robotic octopus works in water and its buoyancy is close to neutral.

In analogy with the biological octopus model, embodied intelligence can be found in the reaching and fetching movements, in locomotion and in swimming.

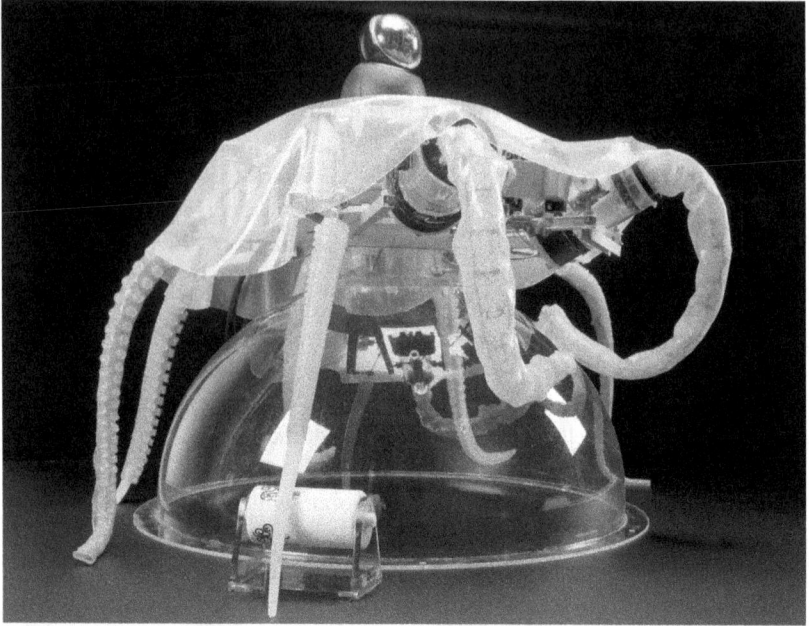

Figure 4.2: A photograph of the octopus robot. Reproduced courtesy of Jennie Hills, London Science Museum.

Wearable Robots with Embodied Intelligence Assisting Human Gait through Symbiotic Interaction

According to L. Pons, 'Wearable Robots (WRs) are systems that assist human motion by extending, complementing, substituting or enhancing motor capabilities while worn by the user'.[21] The key distinctive aspect in WRs is the dual cognitive and physical interaction with humans. Cognitive and physical interactions are strictly interwoven: a perceptual process in the human can be triggered by the physical interaction with the robot, and a cognitive process can lead to a modification of the motion of the robot. Even though both interactions are taken into account in typical design approaches, they are not yet considered in a comprehensive way. For example, WRs for gait restoration are typically considered as advanced assistive tools, with limited autonomy. This perspective is undergoing a major change, as WRs are now more and more intended as *symbionts* with regards to their users.

The term *symbiosis* (from Ancient Greek σύν '*together*' and βίωσις '*living*') has for a long time been used with reference to people living in a community. At the end of the nineteenth century the investigation of nonparasitic interactions involving microbes introduced a new meaning, dealing with the *mutualistic relationship of unlike organisms*. In 1960s, in 'Man–Computer symbiosis', J. C. R. Licklider formulated the visionary concept of human–computer symbiosis, according to which computers and humans would become seamlessly interdependent while sharing common goals.[22] After about fifty years, one may argue that this vision became reality, leading to an ubiquitous diffusion of laptops, tablets, smartphones and many other hand-held devices with advanced computing capabilities. Considering the success story of ICT technologies, the question now being raised is: *which steps are needed to make* human–robot symbiosis *viable* or even *pervasive?*

Human–robot interaction is generally seen as being *symbiotic* whenever a human subject and a (semi)autonomous robot interact with each other to draw mutual benefits by partnering in achieving a *shared goal*. This definition applies well to service robotics, especially in the field of assistance to elderly or disabled people in domestic environments.[23] In these scenarios the common goal for the robots (typically mobile platforms with manipulators or humanoid systems) and the human subject is to enable or facilitate the execution of daily activities. The *mutual benefit* consists in the fact that the human subject profits by an evident physical assistance from the robot, while the robot takes advantage of the observation of the human behaviour to *learn how and when to execute a specific task*.[24]

This learning process, based on observation and cooperation, goes on until the robot becomes fully autonomous and can anticipate the subject's needs. Thus the robot is typically endowed with learning and, more generally, cognitive capabilities and the human–robot interaction is based on a continuous exchange of information between the two agents. This information is related to the way the human subject perceives and gets used to the robot action, and to robot ability in recognizing and understanding the human gestures, intentions, emotions as well as psychophysical state. Besides this kind of information exchange, in some cases, additional sensorial stimuli can be used to further improve cooperation, for example to coordinate/synchronize human and robot mutual actions. In this form of artificial symbiosis the physical/mechanical interaction between the human subject and the robot is not necessary intimate and is usually limited to specific cooperative manipulations.[25]

In *Wearable Robotics* the interaction between the robot and its wearer is inextricably characterized by the *intimate physical interaction*, similarly to what happens in nature between symbiont organisms. *Biological symbiosis* is a complex process where two (or more) independent organisms, *co-evolved* for optimizing their body and behaviour to the co-existence, unconsciously contribute to the *symbiotic system* while pursuing egoistic purposes. Several examples in the biological domain (including humans and their intestinal bacterial flora, needed for digestion) demonstrate that symbiosis is a successful strategy for creating systems which are better than each of the symbionts in terms of survival capability.[26] In a sense, *symbiosis augments both symbionts*.

The effectiveness of biological symbiotic systems suggests that the biological concept of symbiosis could be extended to human–robot interaction contexts where a close physical human–robot interaction is dominant, as in the case of wearable robotics. In this perspective, the design focus is shifted onto the design of proper interaction between the biological and the artificial construct.

A number of issues arise in the design of WRs, mainly related to the need of high levels of efficiency, robustness and safety. Moreover, WRs have to cope with the human body's own dynamics, which is rather complex and influenced by a number of concurrent biomechanical and neurological factors, sometimes not yet completely understood: the human body is an unstructured environment for the WR interacting with it.

Therefore, the fundamental open issue in WR regards the development of a suitable design methodology, where the human body is not considered just as a factor posing a number of design constraints, but rather as an opportunity for a fruitful interaction in both the embodied intelligence and in the symbiosis sense. Such design methodology should be capable of assisting the designers in embedding proper morphological computation capabilities in the devices: the artificial component has to be designed so that the symbiotic system (human + robot) exhibits the desired emerging behaviours.

In a conventional design and optimization loop the dynamics of a machine is given as input to a control module. The designer leads the optimization process, deciding whether to modify mechanics or control, until a target performance is optimized. In a design approach, compatible with the embodied intelligence paradigm and specifically intended for the biomechatronic design of WRs, robot design starts from the co-design and co-optimization of dynamics and control.

The output of this process is a set of detailed dynamical and kinematic requirements (e.g. robot topology and morphology). An approach to implement this design process is not to look for natural solutions to gain inspiration from, as wearable robots have no natural counterpart. Rather, it is useful to get inspiration from the way nature designs organisms. The proposed bioinspired design methodology is indeed based on computational methods, where both morphology and control undergo an artificial co-evolution until a given fitness function is maximized. The objective of the artificial co-evolution process is to assist the design of a WR, which, once coupled to the human body, gives rise to desired emerging dynamic behaviours.

For exemplification purposes, let's take into account a real design case, aiming at the development of a lower-limb WR for gait assistance. The proposed design methodology involves the following steps:

Definition of the search space

In order to point out the optimal design of the WR, the search space must be exhaustively explored. The search space is the set of all plausible WR topologies. The elements in the search space are first enumerated, and then only plausible topologies, i.e. topologies which assure the desired mobility and kinematic compatibility with the human body, are taken into account.[27] Each selected topology constitutes a class of corresponding morphologies, i.e. a set of design possibilities sharing the same topology but exhibiting different morphological properties (i.e. links lengths, inertial properties). Each morphologies set is explored by resorting to numerical means, where the interaction between WRs and the human body is simulated (see next item).

Set-up of simulation tools

A candidate morphology is first selected by simulating its interaction with a passive human body wearing it. In this first step, purely dynamical interaction (active robot – passive human body) is evaluated. In a second step, a refined model of the human body is used, where human adaptation to the robot is taken into account. In particular, this advanced human body model is used to design and optimize the controller of the robot. The models of the human body, of the WR and of the environment are merged together in a simulation environment to search for the optimal control-morphology couple.

Reality check

Since even sophisticated physics-based simulations lack the ability of encompassing the variety of physical phenomena occurring in the real world, which often impact the accuracy of the simulations outcome (*'reality gap'*), before actually fabricating the selected WR, experiments can be performed by rendering the virtual WR using a real robot, i.e. by programming another WR in such a way that it exhibits the same behaviour as the selected WR. In this way, it is possible: (1) to narrow the reality gap, providing feedback on how to improve the simulations; (2) perform experiments with human-in-the-loop, in order to assess the acceptability of the robot and, more generally, all those aspects which depend from human factors, which can be hardly taken into account in simulations with a sufficient level of reliability.

Development of dedicated hardware and software components

The development of a WR exploiting emerging symbiotic interaction with the human body asks for dedicated hardware and software components, including novel actuators suitable for interaction control;[28] novel sensors useful for user's intention detection; novel control strategies seeking symbiotic interaction.

The LENAR (Lower-Extremity Non-Anthropomorphic Robot) is an example of a WR developed according to the methodology above described (Figure 4.3). The LENAR is not anthropomorphic, and its specific kinematics provides a number of advantages, the main ones being:

- *Easier wearability*: small anthropometric changes are intrinsically compensated by the capability of the robot to slightly adapt its configuration; moreover, there is no need to align robots joints to human legs joints.
- *Dynamic advantages*: the heaviest parts (actuators) can be located close to the trunk, thus reducing the oscillating masses, which would have required additional torques for dynamics cancellation.

Open Challenges

Exploiting embodied intelligence in wearable robotics is an effective strategy to build *better* robots. Still, important research challenges, related to the specificity of wearable robots, remain open.

The first research challenge is related to the ineludible complexity related to the proper modeling of the human component – as typical for a genuine biomechatronic, top-down design approach. Human motor strategies affecting body motions and adaptation are *still far from being sufficiently accurate*. Indeed, human models cannot be limited to biomechanics, but should involve *cognitive aspects* (as needed to model human adaptation to external perturbations) and possibly *psychological* aspects (as needed to evaluate, a priori, the acceptability of a given assistive system). Moreover, human models should be flexible enough to cope with inter-subjects differences. The above-mentioned challenges become even more pronounced when the model should account also for specific pathological conditions.

A second research challenge is related to the sensitivity of the model to parameters accounting for the interactions among the robot, the human and the environment (e.g. *feet-ground contact models*). Inaccurate models provide wrong estimations, which, in turn, reverberate on an over-estimation of torques at joints. No doubt, the development of WRs need a basic research effort to develop refined physics-based interaction models to appropriately inform the design process.

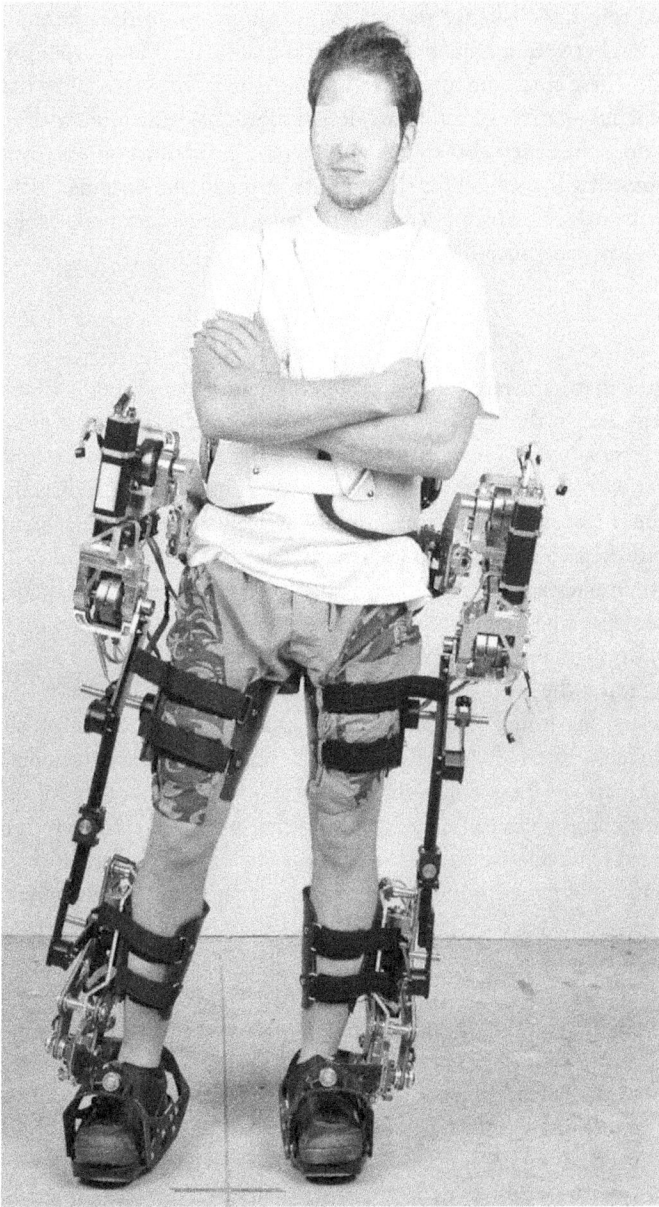

Figure 4.3: The LENAR (Lower-Extremity Non-Anthropomorphic Robot) developed at UCBM for gait assistance and rehabilitation. The robot was designed according to a methodology aiming at eliciting emerging dynamic behaviors from the interaction with the user.

The third research challenge regards the investigation on the optimal level of cooperation between a human designer and a simulation-based tool for design assistance. These emerging tools have an enormous potential to allow the exploration of a huge variety of candidate design solutions compatible with the same specifications, but they also still show inherent limitations on the accuracy of simulations that is essential for the viability of the design outputs. Such limitations can be overcome only if a human designer is able to properly exploit these tools being aware of such limitations.

Conclusions

If we agree that hardware can perform computation, e.g. bring symbiotic benefits to the user in the case of wearable robots, then the rationalist mind–body dualism (*res cogitans – res extensa*) fades out as intelligence becomes embodied. Moreover, we have to accept that computation can occur also without calculation (i.e. without resorting to the evaluation of logical or mathematical expression). Of course, the two implications are strictly interwoven. Computation does not mean, nor implies, calculation, and we can exploit the computational capabilities of the physical world to solve engineering problems through physical models, not only through mathematical models.

Inspired by the view of Konrad Zuse, for whom the physical world is but a computing machine,[29] the environment can be seen as not only a source of uncontrollable external design parameters. Rather, it offers the opportunity to obtain performances (including energy efficiency, capability to self-adapt, simple structures and minimal controllers) that would be extremely complex if approached mathematically. Moreover, some measurable performance improvements can be obtained only if computation is performed physically, through embodied agents. Such benefits include, for example, a zero delay from input acquisition to output generation and a flat spectral response at whatever high input frequency.

From Galileo onwards a phenomenon is not considered to be really known until it is expressed mathematically. But this form of knowledge has its limits, because the mathematical language may become extremely complex in describing several seemingly simple phenomena. The crashing of a drop of rain on the ground, or pedalling on a bicycle or the rippling of a caressed cloth, cannot in fact be easily described analytically, and when they are, solving the resulting equations would require surprisingly high computational capabilities. *Il Saggiatore* was written in 1623. We suffered almost four centuries of Euclideanism, which built a mathematical wall rigidly separating computation (on the human side) from matter (on the nature side).

Certainly mathematics, if not the only instrument, is the best instrument we have *to analyse* the physical world. But we would strongly limit our capabilities if we consider mathematics as the *only* (or the elective) tool *to act* on the world, which is the vocation of engineering. Soft-computing and evolutionary techniques shed new light onto matter and offer insights on how to exploit the embodied intelligence paradigm in the biomechatronic design of advanced systems. All this is not about building intelligent robots, but just useful machines.

5 MANAGING COMPLEXITY: MODEL-BUILDING IN SYSTEMS BIOLOGY AND ITS CHALLENGES FOR PHILOSOPHY OF SCIENCE

Miles MacLeod

In the history of science there have been many conceptual crossings between the *bios* and *téchne* realms, as researchers have sought to use biology to inspire design or represent biology in terms of mechanical models. Philosophers have been comfortable describing the analogical relationships and their limitations between technological devices and biological organisms, and setting up the boundaries between them. However, technology and computational simulation are creating new interactions and fusions between *bios* and *téchne* in emergent quantitative biological fields like synthetic biology, systems biology and bioengineering that extend to practices of investigation themselves. Technology and the quantitative perspective that comes with it have become *means* of structuring both scientific practice and scientific understanding in novel ways that can address biological complexity.

It is these intersections of a more pragmatic engineering mindset with scientific investigation that give systems biology distinct characteristics that are not easily interpretable with current philosophy designed for understanding traditional scientific practice and scientific disciplines. Notions that work well to characterize biology do not necessarily work well to describe this integrated highly mathematical and computational bio-engineering context which is developing everyday novel concepts and practices irreducible to traditional scientific or strictly technological ones.[1] These differences go right to the heart of our understanding of how these new hybrid fields operate on a day-to-day basis to solve problems. Systems biology as such is not 'normal science', or at least not yet, and currently depends on the inherent flexibility and innovation to handle the complex tasks it faces. Analyses of the field do not yet capture the rich interdisciplinary practice this is generating, nor the diversity of model-building strategies within the field.

This chapter will present an overview of some of the specific novel features of systems biology that characterize the current structure and approach of the field to complex model-building tasks. We consider the central role of mesoscopic modelling in practice, as opposed to more traditional top-down/bottom-up modes; the diverse goals systems biologists pursue and the lack of a standardized methodology; and the diversity of epistemic strategies modellers use to extract information from complex systems. Each seems to require a more novel philosophical account than is currently available. They collectively show that the integration between traditional biology and engineering has novel results. Understanding these allows us to understand how the field currently operates to build models, without assessing it according to the standards of more traditional science, like molecular biology, and thus to cultivate a better understanding of skills and interdisciplinary structures that can best favour its success.

Ethnographic Research

My claims here rely on a five-year ethnographic study, performed in collaboration with Professor Nancy Nersessian, of two systems biology labs.[2] It is worthwhile at the start to give a brief account of systems biology research. Our labs self-identify as doing 'integrative systems biology', a form of systems biology that sees its primary methodological aim as the integration of computational and mathematical methods with experimental biology. They focus on modelling metabolic networks and to a lesser extent on gene regulatory networks and inter-cellular interaction. Researchers in ISB are mostly graduate students and mostly come from engineering backgrounds, which seems to be very common in systems biology more widely, although of course modellers from physics, computer science and mathematics also participate. However, as we will see later, aspects of engineering play a part in understanding how the field operates and the novelty of the field's approaches to investigation.

These labs are nonetheless diverse in other ways. The first, Lab G, is composed only of modellers, a proportion of which model biological pathways and the rest that work on generating mathematical methods for parameter-fixing or structure identification. The former set of modellers work in collaboration with molecular biologists from outside the lab on pathway modelling projects. The second, Lab C, includes modellers, experimentalists (trained in molecular biology), and researchers who do both experimentation and modelling, although usually with some collaborative support from molecular biologists from outside the lab for theoretical guidance. The claims about modelling in this chapter derive from our analysis of ethnographic interviews around the model-building processes and practices of the modellers in both these labs.[3] We correlate the results with a broader understanding of the field through communications with our subjects about their perceptions of how the field works and broader knowledge that comes through the published literature. Although of course there are limits to what can be drawn from just two labs, especially given this field is so diverse across many dimensions (including lab organization, research background and methodology), we have been able to extract good indications of what the contrasts are between these labs and others, and how well their experiences correlate with other labs.

The Diversity of Systems Biology and the Philosophical Questions It Raises

Systems biology is a diverse set of practices that cannot be analysed by a single set of goals, or a single set of interdisciplinary relationships. In this section we will illustrate this diversity and the novel issues it raises for those philosophers interested in scientific practice. Some general aspects of the field, however, can be identified as a starting point to investigation of the field. One can best understand what unifies these practices as a shared commitment to model complex biological systems using computational and mathematical resources, and to an often loosely specified idea of a 'systems approach'.[4] The modern incarnation of systems biology, descending as it does from a variety of different roots like dynamical systems theory, cybernetics and smaller-scale modelling in molecular biology, is about twenty years old. It is born of the widespread availability of adequate computational power, developments in mathematical and algorithmic techniques, and the invention of mass data production technologies such as high-throughput data machines that collect dense dynamic information from a system.[5] It must be pointed out that this later feature is somewhat over-estimated as a part of modern systems biology, as it still lacks availability to many systems biologists.

Systems biology of all stripes positions itself against traditional biological fields like molecular biology which systems biologists characterize as applying experimental techniques to measure molecular properties, often in vitro, to discover interactions and to build up small-scale pathways. The need for a 'systems approach' is supported with several philosophical claims in this regard.

In the first place, the properties and biological functions of components are dependent on their part within systems.[6] Because of the complexity of these networks (due to the many interacting components and nonlinearities in the form of feedback relations), these networks need to be investigated *in silico* as well as in the laboratory. Since parts and operations are typically determined and modified within the bounds of large-scale systems, only at this scale can representations generally be accurate and means discovered to reliably control and intervene on systems for, say, medical purposes. Operating with small-scale representations and smaller pathways, as molecular biology has done, risks neglecting many important interactions that determine the dynamics of these pathways.

The aims of systems biologists are usually represented by its practitioners as twofold. In the first place the desire is to have large-scale representations that allow the mathematical computation of the control structure of networks, in principle calculable for individuals by adjusting a system model to fit an individual's parameters. Such specificity will assist in personalizing medicine.[7] Secondly, the hope is that mathematical analysis of network models will reveal design and organization principles that characterize subcomponents.[8] These provide a scaffold for larger system investigation, as well as a basis for evolutionary research and potential synthetic network construction. In many cases systems biologists believe such investigations will lead to the production of a general mathematically-based theory of biological system operation and organization.

However whatever the aims and agendas of systems biologists, almost invariably the model-building problems they face involve significant complexity that makes these agendas very difficult to fulfil. Biological networks not only manifest significant feedback, but chemical elements often play different roles in the same network. This can lead, for instance, to competition effects that add nonlinearity. Data is rarely adequate or of the kind required for modelling the networks in a straightforward way without additional experimentation or the development of computational and mathematical techniques that can estimate the missing information regressively and algorithmically. In the later circumstance computational constraints are almost always in play and affect the moves that systems biologists (who often only rely on personal computers) can make. Even then, the complexity of these networks affects the ability of those systems biologists relying on parameter-fixing techniques to perceive the nature of their parameter-spaces and thus intuit where to search them effectively. All this contributes to the complexity of model-building tasks in systems biology, and we must be wary of gaps in modern practice between the rhetorical goals and philosophical claims of systems biology, and what actually can be achieved at this point in time.

Top-Down or Bottom-Up? The Mesoscopic Alternative

The top-down/bottom-up division in the field has been proposed as the central division structuring systems biology by philosophers and by practitioners.[9] Top-down systems biology imports and adapts concepts and techniques from systems engineering using new high-throughput data technologies to 'reverse engineer' very large-scale system structure based-on the properties and dynamics of the system. Bottom-up systems biology aims to build a systems-level representation by piecing together experimental data and biological knowledge of pathways. Most streams of systems-biology however are neither purely bottom-up nor top-down, but assemble information from different contexts and employ computational methodologies to infer structure where required. This distinction in fact lacks philosophical informativeness since much modelling in systems biology is integrative of both systems and molecular level information. This integration plays a significant role in the ability of systems biologists to find solutions to their complex model-building problems.

Much bottom-up research is neither top-down nor bottom-up but in fact mesoscopic, working from integrated abstract representations of middle-scale networks in order to build out models given cognitive constraints.[10] This type of modelling is a form of middle-out modelling (a term attributed to Brenner by Noble), which provides a platform for modellers to use models as a platform for further investigation.[11] Systems biologists themselves consider it to be a better way of characterizing what they do.[12] As Voit et al. put it, 'If one would survey all computational systems models in biology, published during the past decade, one would find that the vast majority are neither small enough to permit elegant mathematical analyses of organizing principles nor large enough to approach the reality of cell or disease processes with high fidelity'.[13]

In both our labs mesoscopic modelling is a technique researchers rely upon to help break down the complexity of these integrative problem-solving tasks. The technique however places limits on the scale of system modellers can tackle, which limits network size usually to a middle-scale. This size is larger than what molecular biologists typically would investigate or the level at which mathematical analysis of design is possible but typically much less than the large-scale at which biological phenomena likedisease are fully exhibited and controlled. However these limits are also expressed through representational limitations built into these models. The representations developed in the cases we have researched neither fully account for the molecular level nor the systems level, but are an accommodation of each level as shall be detailed below. As such this strategy cannot be represented as integration through sequential top-down and bottom-up steps but is rather an attempt to generate an integrated representation of inter-level relations from the start that help facilitate moves to look both up and down.[14]

Our modellers start with data for the most part acquired from the literature and/or experimental collaborators. They use this to assemble their own pathway representations of the systems they are interested in. This is usually itself a novel result, since these pathway representations aren't normally to be found in the literature and involve modellers making their own inferences, and researching themselves what potential factors are likely to be important in a network's dynamics. At this point they move to building a mathematical representation usually in the form of an ODE model. Even now the data and information available sets limits on what they can model. Data, as expressed above, is almost always inadequate for dynamic models, and often the biochemical processes involved may be too small to be measurable. In lab G these problems are left to parameter-fixing processes, but local parameter-fixing (like regression analysis of particular parameters) always introduces errors. Further global parameter-fixing tasks can't be arbitrarily large given computational constraints, and the greater the complexity of the problem the harder it gets using these processes to find robust solutions. In lab C there are limits to what can be experimentally computed, in terms of time, but also due to the structural limits of experimental equipment and protocols (which are generally limited in what types of interactions they can measure) as just mentioned. In either case modellers work to shrink their parameter spaces by creating abstract representations of their networks. This involves applying sensitivity analysis to pick out only particular parameters as causally dominant, applying biologically plausible or numerically simple values (such as one's and zero's) to other elements or removing them from the model, or using data-fitting regression to calculate the ones for which some data is available. Lastly global parameter-fixing processes will fill in the remaining uncertainties (usually a significant amount), terminating rarely with a best fit, but often with a family of models with similar residual error. Parameter-fixing serves to integrate higher-level knowledge about the system represented in the form of data about system variables of interest. It is these that the modellers want to reproduce, and parameter-fixing serves to find the versions of their network representations that do.

At the same time modellers set limits on the kind of accuracy they are willing to obtain from their results, limiting the subset of relationships for which accuracy is strictly required, or the perturbation range over which robustness is expected. Model-building is thus in practice a highly iterative process of constraining the representation from both ends and also the scale of the model, until an accommodation can be made between an approximately accurate representation for a given purpose and mathematical tractabililty.

The result is an abstract representation that is mesoscopic, not just in terms of scale, but also in the sense of being abstract with respect to the lower-level, and approximate and partial with respect to the higher-level. These models target specifically interlevel relationships, by restricting models to representing just particular relations between higher and lower phenomena. Abstraction and the filtering out of important relationships is assisted by canonical mathematical frameworks like Biochemical Systems Theory, which apply generic abstract ('non-mechanistic' in their language) representations of chemical interactions that are relatively mathematically analyzable and fixable, compared to say Michaelis-Menten representations, but can be fixed to account for most non-linear behaviour possible in metabolic networks.[15] Whether using these frameworks or not, the final models serve to integrate and abstract both lower- and higher-level information.

The integrative process importantly is a process of looking up, down and around.[16] Models are reduced to a tractable core but can then be expanded by building in more information from higher and lower levels. The generic abstract relations of BST (power laws), for instance, can be reformulated with more mechanistic representations, which is considered one of their chief advantages.[17] More elements can be added to pathways. Better data on parameter values can be added. They are also mathematically scale-free, and any sub-network can be re-represented in terms of power laws. As such with a reasonably robust starting model, these steps are simpler than they would otherwise be.

But there is likely even more going on here that is of direct philosophical interest in terms of how these engineers handle complexity. Mesoscale model-ling of the type we have described provides the modeller an integrated multilevel perspective of how higher and lower levels relate in the neighbourhood of solu-tions to the overarching problem. Researchers do not attempt at the outset to provide faithful or accurate detailed representations of either higher or lower-level behaviour but representations that mediate between the two, trading off details and scale in order to construct, at the start, a problem they can reasonably hope to solve that puts them on the path to an overall solution. The process takes the form of a significant problem transformation, taking an impossibly complex opaque problem and generating one that is potentially much more transpar-ent. Finding an initially solvable problem provides potential information about where to search for further improvements to the model, and thus helps struc-ture the search task. By generating a mesoscopic model the causes and effects of changes and additions to the model can be more easily discerned and anticipated. Multilevel problems are usually hugely complex since lower level perturbations to these models don't track linearly to higher level perturbations, particularly with models of metabolic and gene regulatory systems. Cautious modification to the model can be simulated, and the effects tracked and understood. At the same time the model itself provides certain expectations about patterns of cause and effect between the levels researchers can rely upon for their next moves. These models, in other words, bring levels into partial coordination that allows their further coordination to be more easily explored and identified. This gives researchers some understanding and control over how their levels relate. Through such processes we witnessed several times modellers infer the existence of biolog-ical elements in the pathway that weren't known to biologists themselves.

Spelling out what is going on here requires more philosophical development, but whatever is going on is novel in the context of biological investigation, a product of the quantitative skills and concepts engineers bring to debugging complex systems. Importantly, philosophical understanding of these practices has ramifications for how we communicate and educate effective model-building in these complex contexts. As we have observed ourselves, the engineers who enter our labs for graduate study are usually quite confused about their tasks at the start and struggle to develop effective model-building strategies to get a handle on their systems, through for instance mesoscopic modelling, which is a unique skill. Describing it to new modellers could certainly help flatten out the learning curve. Yet mesoscopic modelling both ontologically and epistemically is distinct from the multilevel processes usually observed and discussed by philosophers from other contexts. These discussions for the most part study multilevel model building practices that are sequential processes of adding information from one level and correcting it from another, or where higher and lower level research were carried on at separate times and the integrated.[18] Our results from systems biology suggest, however, that multilevel modelling may depend on more integrative interlevel representations which require a different philosophical account. Having such an account will be of service to the community of systems biologists to have at hand and better guide modellers entering the field.

Diversity of Agendas

Systems biology is usually portrayed in terms of the large-scale representation and design principle goals we represented above, where top-down and bottom-up represent the different methodologies for achieving those goals. But the commitments to these goals tend to vary. As O'Malley and Dupré point out however there is not a well-developed concept of 'system' that all groups and sub-groups share.[19] Furthermore most systems biologists are rather ambivalent towards pursuing a general theory of biological systems and acknowledge that a sufficient class of well-validated and robust models do not yet exist. O'Malley and Dupré describe the current shared commitment of systems biologists as a commitment towards an approach that 'foregrounds mathematical modelling in order to transcend piecemeal analysis'.[20] To get an adequate understanding of the field philosophers have to be careful foregrounding those systems biologists solely that present highly theoretical perspectives and theoretical results, even though of course such results are often of immediate philosophical interest. Using data from our labs we can give an expanded account of the underlying agendas of various laboratories that has particular relevance for lab organization, the scale of systems that labs tackle and the kinds of questions they are interested in. Indeed for a deeper and practical understanding of systems biology the distinction we talk about below is probably more important than top-down and bottom-up.

Our two labs in fact have quite distinct modelling agendas. Lab C, for instance, the lab with the fully equipped wet-lab, is dedicated to a single research issue, which is alive and well in molecular biology and physiology; namely, what role do reductive oxidation processes play in cell signalling? Their overall hypothesis is that it plays critical roles. The lab C PI, C4, sees the role of modelling here to help confirm this hypothesis and map out areas for experimental investigation. Modelling in this respect is a tool for a more sophisticated molecular biology. Here's C4 describing her role.

> *C2 [postdoc] comes from a more traditionalist biology background. And so he gets really into the nitty gritty, he gets really excited about these 2 proteins and wants to investigate them to the nth degree ... I tell him that there a lot of labs out there that can do this much easier than us and that's not what we really specialize in – what can we do that no one else can do – and put it in terms of the context with respect to the rest of the network.*

Modellers like her can supplement traditional molecular biological investigation by providing a systemic context for understanding molecular function.

Lab G, however, is driven by a quite different agenda. The lab G PI, G4, has the explicit goal of advancing the mathematical investigation of biochemical networks. He aims at a sophisticated mathematical theory that will in time produce something akin to laws or at least powerful generalizations about biological networks. Many systems biologists are upfront about seeing systems biology as a potential 'theoretical physics' for biology.[21] They see mathematical and computational development necessary to breaking down the complexity of these systems and analyzing them, a function that goes beyond just modelling systems. Indeed model-building and analysis of increasingly larger systems will only be possible with concurrent mathematical and computational research.

These are two quite different agendas which seat in our two labs quite different practices. These practices trade-off different epistemic preferences and results in an element of tension between the two labs since these preferences are likely to lead to big decisions in terms of how systems biology will organize itself going forward. In the lab G case the lab is organized around developing mathematical skills for model-building and model analysis. Its modellers do no experiments themselves. This forces them to rely on experimental collaborations for data and biological expertise. Unfortunately these relationships are often vexed.[22] Modellers can't get the experimentation or biological advice they need to back up the critical steps they need to advance their model-building processes. There are many reasons why these relationships don't function well, relating to a lack of understanding of each other's practices and goals on both sides. Structurally however the situation is always unfavourable to the modellers. As the lab G PI put it, '*and I still maintain, I've said it for 20 years, you need 10 experimentalists for every modeller*'. This is just not feasible at present and won't ever be unless molecular biologists can be convinced of the necessity of systems biology and there are significant changes in the structural organization of molecular biology.[23] As such a significant proportion of research in lab G is arguably orientated towards reducing reliance on experimentation through the development for instance of better algorithmic parameter-fixing techniques. The resulting models are however often treated extremely suspiciously by molecular biologists who don't understand why they should treat abstraction and approximation processes as reliable (like BST or parameter-fixing). Biologists are in general very skeptical of any claims drawn from limited sets of data. Further molecular behaviour is substantially variable across contexts and as they see it systems biologists frequently neglect this when putting together large scale models from cobbled together data sets as they seem often forced to do.

Lab C on the other hand works much closer to experimentation, avoiding the problems of collaboration. Modellers do their own experimentation when required. They can confirm their results as they go and discover areas that require experimental investigation. Importantly they have less need for advanced parameter-fixing techniques and often end up with only a small collection of parameters that need fixing. The rest they can measure. Even more importantly, as I and Nancy Nersessian argue, this bimodal strategy creates extra affordance for localizing and resolving errors in models, by allowing modelers to move strategically back and forth between a model and experiments.[24] This 'bimodal' style of practice allows modelers in turn to build more 'mechanistic' representations of networks using for instance Michaelis-Menten interactions, which biologists tend to appreciate and understand more than say power-law representations. Their models have as a result a better chance of being published in mainstream biochemical literature.

But while the bimodal style seems to have distinct advantages it also has critical trade-offs from the perspective of lab G. It restricts the scale of the system that can be handled given the investment in experimental research and lack of development of mathematical skills by the researchers that do it. If an aim of systems biology is to apply mathematics as the key to getting control of systems, and system control is only effective at a large scale, then lab C work seems to fall quite short of that agenda. Further lab C work doesn't explicitly promote the value of mathematical analysis of systems. Indeed using representations like Michaelis-Menten, which are hard to mathematically analyze at large scales, lab C seems to work against theoretical goals and deeper theoretical realizations about the structure of biological systems. According to G4 *'the models that are being developed in C4's lab are ... by and large, off-the-shelf type modelling approaches that are, not always, but that are often rather simplistic, I'd say'*. Here G4 was citing the lack of mathematical sophistication of modelling techniques or analysis in lab C and the use of rather common *'superficial'* mechanistic models of interactions such as Michaelis-Menten models.

The relatively simple unimodal-bimodal division in our labs throws up all sorts of deep questions about what systems biology is for and how best to do it that goes beyond top-down and bottom-up. Systems biologists are starting to come down forcefully on one side or the other and this of course effects how they train their graduates and structure their labs. Other labs, both top-down and bottom-up, experiment with mixing experimenters and modellers together (see for instance the Lauffenberger lab at MIT), but there are still basic affordances to being both a modeller and experimenter that can't necessarily be achieved this way. This is a big issue that likely affects how molecular biologists perceive systems biology and its competencies. The two agendas we have detailed here are however unlikely to be the only two agendas in systems biology. Diversity runs deep. For instance there are groups of highly computational systems biologists (such as those modelling data for medical science) that are very pragmatic in their attitudes (not much interested in theory) and also not particularly close to molecular biology. They are results-driven big data biologists. Understanding these agendas will be extremely important to correctly philosophically interpreting the field and the epistemic positions underlying it. This will help those outside the field, like molecular biologists or clinicians (as potential collaborators), understand what's going on, but it will also help cultivate within the field a better understanding of the skills and organizational options available, and how they might fit the immediate modelling goals a particular researcher might have.

Diversity of Model-Building Strategies

So-called 'normal science', the science of established disciplines, is science that works from established methods and task routines.[25] Molecular biology has well established experimental artefacts and tools appropriate for experimental investigation of molecular subjects that prescribe particular protocols for their use and a set of resources which experimenters learn for unpacking what to try when things go wrong. A more theoretical discipline has well established bodies of theory out of which models can be formulated for particular circumstances. Systems biology is however, currently at least, unlike either of these. The field has not yet achieved standardized procedures and theories that can address especially all the different kinds of networks and data situations modellers find themselves with. This means there's no reliable general theory yet of biological systems or specific classes of system, but also no theory that can inform modellers how to model particular systems that if followed will reliably produce good outcomes (compare for instance the canonical theory in physics that modelling fluids start almost always from the Navier–Stokes equations).

Canonical mathematical frameworks do of course theorize about the behaviours (or at least ranges of them) of biochemical systems, but these are not always appropriate given variable data situations and network behaviour. Modellers may have only steady state/equilibrium data; they may have incomplete time series data; they may only have *in vitro* data. Almost always modellers need to tailor their approach to fit their data situation. As G16, a graduate researcher who came to lab G from telecommunications engineering told us:

> when I talk to G10, uh his project, like his kind of data are different. Like he has it for like different genes knockouts and then, um more of steady state data. And then like G5's data are different. His are [not time series] ... and then you could [get] creative with it, like you could say, um I've tried different things and then it took me a while to realize that's not the way ... that's not what I can use. Like G10 could use it for his project because of this and that. For me, it's not going to work because I have this dynamic data, which is different.

This was contrary to her expectations and certainly it is for many modellers we have talked to that expect a more programmatic way of building models, but end up spending much time just figuring what they can get out of their systems with the data they have. This requires for instance choosing a modelling framework that best seems to fit what is possible with the data, finding ways to extrapolate data, making simplifying assumptions about network structure or parameter values to fit what is possible with the data, and modifying the modelling framework mathematically.

Most problem-solving work is as a result a highly iterative, not linear, process of going back and forth between the model and the data, or the model and experimentation, and between model-building stages (pathway representation, mathematical representation and parameter-fixing techniques), while simplifying each in order to make other steps workable. Modellers are often walking a very fine-line between finding a problem that can be solved given the governing constraints on collaborative relationships with experimenters, and getting a robust representation.[26] Further there are many modelling frameworks and choices that modellers can make, such as choices about what parts of pathways to represent and interactions to model; how to mathematically represent the interactions; which computer programs to model in; which data sets to rely on and so on. As G16 explained the problem, 'when you say modelling, it's very broad ... no there are not routines. In each case, you see what kind of data you have and what you could do actually with them.'

In this context mesoscopic modelling is part of this very process of finding a manageable tractable problem. What we have discovered is that modellers are in fact highly innovative in terms of how they find that accommodation between higher and lower-levels, and the governing constraints.[27] They make clever mathematical inferences on the spot that sometimes lead to the discovery of new chemical elements in a pathway. They develop their own novel algorithmic techniques and mathematical techniques for analysing their systems, knowing themselves where their representations are likely to be robust and informative, and where they are not. In the language of Weisberg and Muldoon systems biologists are methodological *mavericks* rather than *followers*, searching a wide space of the epistemic landscape that describe the methodological choices in systems biology, and experimenting with different strategies.[28] This makes sense where that landscape is either complex, or at least uncertain. Rationally, allowing their researchers to roam like this, as our lab PIs at least do, increases the chances of some strong and general methodological procedures being found. Still at present these don't exist in our labs, leaving graduate research with cognitively difficult search tasks that requires much complex work on their behalf.

Our two labs themselves only represent a small slice of what is going in a field that ranges from highly computational and mathematical labs which barely collaborate, to highly experimental alternatives. Labs may structure their whole methodological approach around single collaborative relationships and the type of data they have, allowing the development of very specific but effective resources. The whole field in this sense is itself experimental. It is experimenting not just with algorithms but also differently organized collaborative relationships and experimental training. No one can tell at this point which methodological strategies will pan out as most effective for any given set of goals, or the overall goal of mathematically modelling biochemical networks.

Philosophers have long focused on the science of disciplines which of course at least approximates normal science (although probably more loosely than we have realized). There are suggestive reasons for this. Such science is science with established methodologies and theories that can be written down, interpreted, and analyzed. Textbooks may reliably map the core knowledge and methods of these fields. Published papers insofar as they employ these methods, can be reliably understood in terms of how conclusions are reached and what methods and concepts they rely on. But systems biology is not one of these, and in fact that's what makes it particularly interesting. Systems biology exposes how innovation works and what the human mind is capable of. It tackles complexity head on without standardized models or widely applicable canonical practices. In this sense too, by not yet canonizing a set of mathematical abstractions, it will remain genuinely interdisciplinary necessarily relying on experimentation and experimental expertise to get its models to work and fit the data. As we will suggest it is the engineering mindset, a concern with practical results and willingness to bootstrap problems, that is driving this. Philosophically however the issues such a field raises and how they work requires different resources than the ones we are collectively used to. We need to move away from the implicit generalizations one invokes when studying theories or methods 'of a field or discipline' and pay attention to the individual practices, and all the factors that go into solving a problem – cognitive and collaborative (social), mathematical and experimental – in systems biology. Only then can we begin to generate a logic of what's going on which we do for instance when we rationalize methodological variability in systems biology as a way to search across a complex epistemic landscape.

Explanation, Understanding and Epistemic Strategy

Finally, this diversity and variation from more disciplinary science extends to the epistemic strategies systems biologists use to establish claims about their systems and the kinds of information they try to derive from their systems. Here as elsewhere systems biologists are innovative in the face of complexity. In the first place generating mechanistic explanations of phenomena may not always be the best framework for characterizing the epistemic goals and strategies of systems biologists, even those close to a bottom-up stream, such as those in our labs. Philosophers of course have now well noted the importance of mechanistic explanation (ME) for encompassing much of what researchers in the life sciences do.[29] This was an important shift away from notions of explanation inherited from logical positivism that had laws or theories specifically in mind. The notion of mechanistic explanation has recently been extended, using the case of systems biology in fact, to cover dynamic systems. In these accounts a dynamic mechanistic explanation (DME) demonstrates how parts and operations are 'orchaestrated' through their organization to produce particular phenomena.[30] Mathematics and computational simulation is usually essential to this.[31] Implicitly the concepts of ME and DME set standards of accuracy on the correspondence of parts, relations and their organization that help guarantee representational and predictive accuracy. Mechanistic explanation with systems can be considered an appropriate normative goal for modellers, but in these highly complex model-building contexts, this may often be unattainable.

As I mentioned above many systems biologists in the course of model-building revise their goals and epistemic strategies to establish their claims based on the data they have and what they think they can get out of it. In lab C options are different because they can produce their own data and can modify their aims to stick within the realms of what they can mechanistically explain. Their agenda is less orientated to tackling up front large scale systems. Lab G participants on the other hand have various abstractive mathematical skills to work around the data problem. What we have found is that lab G participants certainly use mechanistic representations (at least partial ones) but often develop goals that are not explicitly explanatory (in the mechanistic sense) but rather are tuned to extract particular kinds of information they can get from their system that nonetheless advance a systems account and provide some kind of understanding of what is happening in their systems. In the process however information about the 'orchaestration' is lost or obscured beneath layers of abstraction and simplification.

For instance G10, one researcher, refined his goals to mapping the mathematical relationships between particular variables in a system relevant to controlling an important output. In this case he wanted to learn exactly how to manipulate certain enzymes to reduce the amount of woody biomass generated in lignin synthesis. This mass interferes with effective biofuel production. In the case of G12 one of her goals was to form a reasonable hypothesis about the manner in which a certain enzyme called Nox1 regulated inflammatory processes in the body. They constructed pathway representations of the lignin synthesis and inflammation systems. These pathways count as at least as partial mechanistic representations. Neither had much data to go on, but had large networks to deal with. Each employed layers of simplification and abstraction to shrink and simplify substantially the parameter spaces of their models through numerous techniques. G10 used flux balance analysis to linearize his system around steady state in order to estimate many parameters, dumping information about nonlinear interactions in his system in the process. He then used sensitivity analysis of the dynamics to restrict his system representation to only the most significant variables, before globally fixing the remaining parameters. G12 made similar moves, including setting many less important parameters to 0's and 1's to simplify the parameter-fixing mathematics. In both cases errors were introduced with the confidence that the fixing process could nonetheless compensate and find stable approximate results.

However in neither case could either hope to derive a single best fit. Instead, both fashioned arguments based on multiple models with the same residual error. For instance G10 found 5 optimized models that performed similarly over tests with the same residual error. None likely represented the global optimal but the fact that he had five well performing models that converged on a similar mathematical relationship for his target variables gave him confidence to assert that he had correctly found that relationship. G12 had a tougher task. She had to sort through the numerous possible structural relations between Nox1 and elements of the inflammation network. She did by developing her own model-based argument. She had a model of the overall system again built with limited data, with which to test her hypotheses. She tested the alternative roles of Nox1 according to which gave the best performance for the system model across a large multivariate parameter space (using Monte-Carlo). The best performing alternatives she argued gave the most likely relations of Nox1 within the network (i.e. which elements it was interacting with). She didn't present an integrated mechanistic model of the inflammation system as her final result, but had a platform with this testing to reason about the likely role of Nox1 regulating inflammation. In both cases G10 and G12 created their own epistemic strategies to support (however weakly) particular claims about particular information they derived from their systems.

The process however of getting these results arguably washed out their ability to mechanistically explain what was happening using their system models, at least according to the DME account of mechanistic explanation. Parameter-fixing is a topic still under-discussed in philosophical circles. Yet in systems biology it is a particularly significant issue, and lab G devotes a significant portion of its resources to improving it. In first place simplifying parameter spaces distorts the values parameters will be assigned since various parameters are left out or given highly approximate or dummy values, in the hope that the fixing process will have flexibility to compensate. Global fixing processes don't fuss about the accuracy of one parameter: but combinations that render the overall model accurate. The result is often systematic distortion. Parameters often compensate for the errors of one another and many parameterizations may yield exactly the same residual error. Actual global optimal solutions are rarely found, and there may not be an opportunity to have the parameters checked experimentally. Modellers have to go with what they get and turn it into an argument of the kind G10 and G12 used above. The result is that although the parts may to a large degree correspond through pathway models the interactions/relations will diverge since the parameters of these interactions are often inaccurate. As a result the relation between the final mathematical representation and many aspects of the underlying mechanism are often obscured between these layers of abstraction and simplification.

G4, the lab G PI, labelled the content of these mathematical models 'somewhere in between' something explanatory and something associational or phenomenalistic, preferring to see these models as adding specific new causal information about a system but not an overall explanation of system function. Indeed G10 and G12 would certainly claim to have derived some form of understanding from their systems with the specific information they obtained.

> So, if you knock down just one enzyme, you could get a 25 per cent decrease in the S/G ratio. But if you're allowed to change the expression of as many as three enzymes, you could get a 38 per cent decrease in the S/G ratio, which is huge compared to the natural variation that is about 5 to 10 per cent. (G10).

G10's novel understanding thus came in the form of some understanding of the mathematical relations between particular variables relevant to control, in line with the concept of a 'systems-level understanding' which often seems to contrast mechanistic accounts and system-level accounts.[32] G12 assembled claims about the regulatory function of one element within her system, adding certainly some potential causal-explanatory information biologists were unaware of, and hence helping establish the value of mathematical model for providing useful hypotheses about network structure.

There is no doubt mechanistic explanation is a normative standard for systems biology. DMEs guarantee accuracy for the particular goals systems biologists take on, and much better epistemic strategies for arguing for their conclusions, than these more abstract model-based arguments based on multiple models and parameter spaces. Very good projects that run over a long time with good data and biological input often achieve these aims. But if philosophers focus on mechanistic explanations they miss the creativity of these epistemic strategies and the innovative ways systems biologists can develop their goals to achieve mathematical tractability with extremely challenging problems.

It should be mentioned finally that underlying this discussion are the standards the fields themselves set on what counts as 'mechanistic'. Labelling what systems biologists do as mechanistic explanation cuts across their own perceptions of what's going on. For the most part systems biologists particularly in lab G step back from claiming what they do is mechanistic. They are aware that their abstractive and simplification techniques render such a claim problematic. Molecular biologists according to systems biologists give mechanistic accounts by using for instance Michaelis–Menten or other 'mechanistic' representations of interactions. Of course systems biologists don't necessarily have the philosophical resources to argue one way or the other, and certainly simply employing abstractions of interactions shouldn't necessarily deny them the ability to claim to give mechanistic explanations. Both Brigandt, and Levy and Bechtel, have shown that abstraction can in fact serve to draw out relevant mechanistic detail in order to make mechanistic explanation possible.[33] This is however not the abstraction and simplification of parameter-fixing processes described above however, which washes out that detail for the sake of tractability on fitting the data. Nonetheless the reasons for systems biologists thinking this is important draws on deeper facts about what they see themselves doing. These reasons require philosophical investigation about the meaning of a 'systems-level understanding' which is often raised as a justification for abstraction and simplification without much explication of what this understanding consists of. The relations between mechanism, explanation and understanding are complex in these contexts and need careful untangling and more nuanced philosophical accounts. While we can be confident in the normative value of mechanistic explanations, we must be aware that the circumstances of complexity and interdisciplinarity generate divergences from this standard for the sake of mathematical and computational tractability.

The Pragmatic Engineering Mindset: Getting to the Heart of Systems Biology

Remarkably, and fascinatingly, many of these features we have discussed above have something to do with engineering. They have something to do with *téchne*, or at least a particular kind of technologically orientated practice, that is currently entering investigative science. By this statement I do not mean, however, the way in which *téchne* is entering say medicine in the form biomedical engineering or synthetic biology, where clearly the 'engineering' task is to build with new materials and learn something about phenomena in the process. Nor is the relationship between systems biology and engineering to be contained within the fact that systems biology itself studies systems and can be analogized to systems engineering. Certainly some mathematical techniques carry over through this pathway, but not so many engineers in systems biology are fully trained systems engineers. Systems biology is engineering applied in a more subtle way than that, and some of that has to do with the more pragmatic attitude engineers have towards problem solving that can be set in contrast to what say molecular biologists do.

Engineering is often associated with the ability to 'bootstrap', to build in techniques from a variety of sources in order to fashion solutions, rather than to rigorously solve equations or run experimental protocols. There certainly should be a reason why in both our labs engineers are much preferred to other, even mathematical, scientists. In fact the need for the pragmatism and flexibility we associate with engineering is built into the fabric of our labs. Obtaining robust results is stressed over pursuing particular models or applying particular theory. Any kind of model can be valued for this task for the lab PI. In G16's words,

> It's just ... solving the problem. Doesn't matter how. Like figuring out ... getting information from the data ... and the other thing is like changing the system so that you get the effects you want to have ... And then whatever method you want to use that's fine with him..G4 [the lab PI] thinks that if you like get certain results with that – that's enough for a PhD. 'Cause that means you understood the systems.

We describe the underlying methodology associated with systems biology as 'adaptive problem solving'.[34] Systems biologists assemble data, as well as mathematical and computational tools as they go, through a significant trial and error process, before they can find a way to negotiate their problems.[35] Engineers of course have broad knowledge about pragmatic techniques and familiarity with using them. This includes knowledge of parameter-fixing and other numerous approximation techniques such as sensitivity analysis that makes them well suited for this kind of research. They are used to 'satisficing' with phenomenalistic rather than mechanistic results, or something in between. They are also used to debugging. Wimsatt provides an excellent account himself of the engineering approaches to science, the skills of which are well-adapted specifically to handling complexity.[36] Importantly the pragmatism of engineers predisposes them more than traditional disciplinary scientists to interdisciplinary research. Although there is no doubt it is still very difficult for engineers to interact with experimenters, they show a propensity to be willing to learn experimentation when required, or otherwise learn over time how to interact with experimenters by building up some 'interactive expertise' of the possibilities and constraints of experimentation they can build into their own work.[37] This we have observed at least in certain cases, and have tried to encourage ourselves by having the lab G PI place his modellers in experimental labs for short periods. Engineers are perhaps simply less encumbered with disciplinary identities and exhibit the kind cognitive flexibility to engage with others from unfamiliar backgrounds.

Conclusion

Insofar, then, as methodological and epistemic diversity are a central feature of systems biology one can plausibly perceive a pragmatic engineering mindset underlying it. An engineering mindset does not however prescribe, as I have endeavoured to illustrate, here a form of normal science. The *téchne* enters the investigation of biological phenomena in a novel way, just as it does in synthetic biology, but with its own features in this instance, related less to a material dimension than a mental or cognitive one. With this in mind our philosophical accounts of what is going in these cases needs to understand that the logic of this research might be different from those we are used to. Trying to deal with this kind of network complexity, as these engineers and other quantitative researchers seem at least at this stage determined to do, requires its own logic that is driven by cognitive, computational and interdisciplinary constraints, and ultimately uncertainty about the most effective methods to tackle these networks. As such philosophers need a philosophy of scientific practice addressed to understanding how methods and methodological diversity trade-off the relative certainty of disciplinary standards and methods in the field for pragmatic approaches to handling and investigating complexity. In turn we can begin to cultivate awareness amongst researchers in the field itself and its collaborative neighbours of the various aims and goals of systems biology, and the various skills and interdisciplinary structures that help cultivate effective research in this different kind of problem-solving environment.

6 STRATIFICATION AND BIOMEDICINE: HOW PHILOSOPHY STEMS FROM MEDICINE AND BIOTECHNOLOGY

F. Boem, G. Boniolo and Z. Pavelka[1]

Recent advances in molecular biology and biotechnology are drastically changing our perceptions of disease, diagnosis and therapy. In this reshaping of medical activities, molecular and personalized medicine, with their biotech tools, are emerging as an indispensable bridge between basic research in molecular biology and clinical practice.

As usually happens in the history of science, this innovative turn in medicine has also begun spurring new philosophical analyses. In our contribution, we cannot discuss all the scientific aspects and the totality of the humanistic reflections they have stimulated. We will focus our attention on classification, or, more precisely, on *stratification* of diseases, therapies and patients and on what such a stratification implies from the philosophical point of view.

In what follows, we will start describing the topic of stratification, its novelties and its impact on the way of treating patients in clinical contexts ('Stratification Medicine'). Then we will begin showing how the stratification turn has relevant philosophical implications. In particular, we will show how it is changing our way of defining and classifying diseases ('The Philosophy Within'). This is an ontological topic strictly linked with the so-called bio-ontologies, which are a computational approach by means of which we integrate data coming from different sources, especially coming from molecular work, with clinical needs.

Stratification Medicine

Current clinical practice, in the eyes of the proponents of molecular medicine, suffers from two major shortcomings.[2] *First*, the assessment of the patient on which the diagnosis and therapy are based is too rough. It does not do justice to the great heterogeneity of causes lying behind seemingly uniform presentations of symptoms and disease patterns. Additionally, it disregards individual particularities in metabolism and physiology that significantly affect the organism's drug response. *Second*, current medicine is reactive. It comes into action when the disease has already manifested itself in the patient. It focuses on 'fighting' disease instead of maintaining health. In contrast, according to the personalized approach, we should provide 'the right medicine to the right person at the right time' in which 'the right time' is ideally long before the noticeable manifestation of a disease.[3]

Of central importance in this endeavour is the identification and validation of *molecular biomarkers*, which are, as defined by the *Biomarkers and Surrogate Endpoint Working Group* of the American National Institute of Health, '[characteristics that are] objectively measured and evaluated as an indicator of normal biological processes, pathogenic processes, or pharmacologic responses to a therapeutic intervention'. These are established 'based on epidemiological, therapeutic, pathophysiological or other scientific evidence'.[4]

In the literature we can find different distinctions among the biomarkers, depending on the needs of the researchers or clinicians that use them. For example, there is a distinction between *qualitative measures* (e.g. the absence or presence of a certain trait or mutation, or the presentations of different phenotypes) and *quantitative measures* (e.g. staining intensity, measurement of serum levels).[5] One can also find the distinction between *prognostic markers*, used to predict the course of diseases; *diagnostic markers*, adopted to tell apart subgroups of diseases; and *predictive markers*, which allow to calculate the risk of developing a disease or to foretell the reaction to a specific treatment.[6] Others differentiate between *trait markers*, indicating a predisposition for a certain disease; and *state markers*, which are measured during the course of a disease and that inform clinicians about the progression.[7] However classified or measured, biomarkers are considered to be the crucial references when trying to answer questions such as: who is sick? What disease is it? Who will develop a disease? Who should be treated, and with what? How does the patient react to the treatment? Did the treatment work by restoring health?[8]

The vision of both molecular and personalized medicine is both inspired by and relies heavily on the conceptual and technological advances in molecular biology. It is hoped to be the return of the heavy investments in genomics and post-genomics. Consequently, the focus lies on a specific type of biomarker: genes and gene products. Furthermore, with the advances in sequencing and data analysis, these genes and gene products can be assessed in huge quantities. Instead of just taking into account candidate genes or levels of a certain enzyme, there is the opportunity to analyse whole signatures of these molecules and integrate gene expression data with mutation analysis, metabolic and protein profile.

In summary, molecular biomarkers of diseases, assessed singly or as 'signature', promise to revolutionize medicine. And they do this, *first*, by vastly improving our abilities to treat disease through enabling us to sort patients according to their response to treatment; *second*, by vastly improving our abilities to prevent disease by means of enabling us to assess an individual's disease risk and recognize pathological processes long before the illness manifests itself. Let us examine both in turn.

Stratification and Pharmacogenomics

No drug shows the same efficacy, side effects or toxicity in every patient it is prescribed to. Patients can roughly be categorized into *responders*, those who actually benefit from the drug; *non-responders*, those in whom the active ingredients of the drug do not work, however they are still susceptible for the side effects; *adverse-responders*, those who might be more sensitive to toxic effects or show allergic reactions but still might in addition benefit from the treatment.[9]

This divergence in drug response has many potential causes. Some are related to drug interactions with nutritional components, or to the way the drug is metabolized by the liver and kidneys. In particular, inherited genetic polymorphisms in drug metabolizing enzymes also translate into inherited differences in drug sensitivity. Patients with an inactivating polymorphism in the enzyme dihydropyrimidine dehydrogenase, for example, are less able to metabolize fluorouracil, a chemotherapy drug. This inability results in accumulation of fluorouracil in severely toxic concentrations. Polymorphisms in genes coding for cell membrane transporters are also likely to influence drug efficacy.[10]

Another potential cause for differential drug response can be insufficient knowledge about the condition treated. The seemingly homogenous patient population that is taking the drug due to the same presentation of symptoms, is in fact a heterogeneous mix of populations with distinct molecular features which should be targeted by specific drugs. This is thought to be the case in many complex diseases, for example depression or cancer.

The term *patient stratification* when used in combination with personalized medicine almost exclusively refers to the division of a group of patients into different sub-groups regarding their response to a specific drug on the basis of molecular markers like genes and gene products. The identification of carriers of variants indicative for differential drug response is obtainable by specific tests, which accompanies and specifies the diagnosis and the results of which are expected to help picking most promising treatment. One example is given by the testing for K-ras mutation status in patients with non-small-cell lung cancer which allows to exclude the carriers of mutations from cetuximab treatment.[11]

It is to note that in the literature the benefits in terms of wellbeing of patients are strongly entangled with promises of economic benefits. Saving patients from the harm of adverse reactions and pointlessly suffering the side effects of non-effective drugs are mentioned in the same breath as the reduction of expenses for private or governmental health care providers, since it is worthless to administrate drugs to patients in which they either do not have any positive effect or do have a toxic effect. Furthermore, the possibility of identifying sub-populations of patients that react positively to a new drug is thought not only to benefit these patients but also the pharmaceutical companies, as it would make the drug development process much more efficient and the approval of new therapies faster.

Predicting drug response based on molecular markers is especially sought after in the field of complex diseases, most prominently in the oncological field, where the responses to treatment differ widely among the treated population. Indeed, it is exactly in oncology where a great number of new therapies and patient stratification technologies are developed.

Stratification, Risk Prediction and Disease Prevention

While *pharmacogenomics* is concerned with identifying genetic variants that allow the prediction of drug response, *genetic epidemiology* is investigating the correlation of genetic variation with the *risk* of developing a certain disease.

Risk, in the epidemiological literature, is broadly used to refer to a probability of a certain outcome within a population of subjects. This outcome can be unwelcome, like developing a disease or death, but also welcome like immunity or recovery. A *risk factor* is a measurable variable for each subject of a specified population, which can be used to divide the population into groups of high or low risk.[12] Risk factors are mostly found by statistical correlation and are usually subdivided roughly in two categories; *risk markers*, which cannot be manipulated but show a statistical association with preceding outcomes; age at menarche, for example, is a risk marker associated with type 2 diabetes;[13] *causal risk factors*, whose manipulation changes the probability of the outcome, that is, of developing a disease. It is to note that genetic variants that can indicate the increase or decrease of probability of developing a certain disease at some time in life are usually called *genetic risk factors*, even though *risk marker* would be more correct as there is currently no way of manipulating the genomic makeup of the individuals.

There are several approaches to find genetic risk factors for a disease or disease trait for which heritability has been established in twins, families, or population-based studies.[14] *Candidate genes* are suggested by research findings on the underlying biology of the disease, or are known gene variants, that are associated with familiar forms of common, complex diseases like Alzheimer's disease, heart disease or colorectal cancer.[15] In Alzheimer's Disease (AD), for example, characteristic aggregates of peptides (the so-called plaques) can be found in patients' brain. They contain mainly ß-amyloid peptides and the apolipoprotein E (APOE), binding tightly to the ß-amyloid. In case-control studies members of families with late-onset AD were genotyped and an association was found between the APOE ε4 genetic variant and disease occurrence. To this day, APOE ε4 remains the only conclusive risk factor for late onset AD.[16]

Genome wide association studies aim to detect association between one or more genetic variant and a disease or disease trait across the whole population. There is no particular need to take note of the underlying pathogenic pathways, since these associations are statistical links established in case-control studies, in which the diseased population shows a high incidence of a certain genetic variant while the control population does not.

Although enormous efforts have been made and are still on-going conclusive genetic/genomic risk factors for complex, multifactorial diseases like depression, diabetes or heart disease have not yet been established. Moreover, attention has also to be paid to the potential source of bias that arises if the disease and control population share a different heritage. The term *population stratification* used in the context of genome wide association studies refers to an underlying structure of the population used in case-control studies. This can happen if the control population is composed of individuals sharing a heritage different from the case-population. That is, an allelic homogeneity between the two populations has to be at the basis of a non biased study.

Stratification of populations on risk factors and biomarkers for therapeutic intervention is sometimes argued to be only the first step towards 'truly' personalized medicine, which will be achieved by continuous monitoring of a large number physiological and molecular parameters in the single patient, thus creating both an individual baseline-health and the possibility to detect deviations early.[17]

Another possible future avenue consists in the creation of 'patient avatars', that is, a representation of an individual patient in the laboratory on which diagnostic tests can be performed and several treatment options tested. Examples are tissue cultures of patient derived induced pluripotent stem cells;[18] xenograft mice, which are mice implanted with a patient's tumor; or 'patients on a chip' which are systems of microfluidic devices containing patient cells form different vital organs.

The Philosophy Within: Epistemology, Ontology and Ontologies

As we have shown in the previous sections, *stratified medicine* heavily relies on new technical and methodological innovations. In particular contrary to intuition, stratification is pursued and obtained in a bottom-up way. This is because rather than imposing groups and subgroups from a previous theoretical level, such a partitioning of a human population arises from the bottom, that is, from data collected through large-scale analyses. These classification procedures are dissecting human population in ways that were simply unknown just a few years ago. People can now be grouped according to similarities, previously hidden or ignored, pertaining to the molecular details and mechanisms of the genome's structure and behaviour. From a philosophical perspective such a situation, characterized by an enormous amount of data, does not simply state the many classifications which are 'in nature', but rather how much classifications depend on the methods of investigation and, therefore, by the technological innovations permitting them. From a technical point of view, these types of analyses are now possible especially due to new computational tools.

However, this new perspective is not just methodological. Precisely this possibility of producing and analysing an enormous amount of data constitutes the theoretical premises for a variety of claims on what someone (maybe with too much emphasis and a not too deep knowledge in the history of science) calls a new way of doing science. *Data-driven science* is an expression now widely accepted to describe those scientific settings in which the general aim, rather than hypotheses testing, is generating massive amounts of data in order to discover patterns that then would constitute the core matter on which to start the experimental work. Accordingly it is possible to understand the rapid increase of the so called *omics*. All the *omics* pursue the aim of producing, as global as possible, specification and quantification of molecular components in order to provide a general understanding of biological factors and interactors within an organism or among different organisms.

For example, as already mentioned, *pharmacogenomics* deals with the identification of genetic variants within a population in order to predict specific drug response. By increasing the complexity, a more comprehensive understanding of particular genomic profiles associated with specific lifestyles can contribute to reshape our understanding and definition of diseases. The ontological dimension here lies precisely in the fact that a discipline like pharmacogenomics is 'cutting the world' so that, at the end, we can find the personal/unique signatures which should, in theory, define the individual. In this journey from 'types' to 'tokens', let us consider for example the case of *leukaemia*. The term has been originated to describe a type of cancer affecting the blood or the bone marrow and, thus, has been born as a way to address a specific kind of disease. Today, on the other hand, 'leukaemia' is a term which labels a spectrum of different conditions. In the past leukaemias have been classified into acute or chronic ones, depending on how rapidly the disease develops. Other classifications have involved the types of cells which are affected. However till the comprehension on how different cell types could be differently impacted, the granularity of such a classification did not take into account the dimension now unveiled by the *omics* approaches. The possibility of 'looking at the whole' by dissecting the disease at its molecular level opened the chance to deconstruct the disease itself as a single entity and to produce a 'thriving jungle' of different conditions determined by distinct *omic profiles*.

Such a situation is paradigmatic to show how much a genuine philosophical analysis can actually stem from the very practice of science. In other words how should we interpret this change promoted by stratified medicine? Is it just a methodological innovation or truly a new way of doing science? What are the implications for medical practise? To address these questions the philosopher (but even the scientist) should try to avoid to dictate his/her conceptual categories. On the contrary, the purpose of an ontological/philosophical analysis pertaining to this context, is not to impose an *a priori* ontology to constrain the scientific work. It would be rather the opposite, that is, to highlight the hidden ontological assumptions in that context. In other words this situation offers a hint on the fact that there is the genuine possibility of doing philosophy *from* and *for* science and not just *of* science.

Ontological Categories and Epistemic Framework

The proliferation of groups and subgroups produced by *omics* in stratified medicine should not be interpreted as a multiplication of ontological categories. Indeed a modification or a change of ontological commitment does not imply, *per se*, a metaphysical standpoint. Rather, it is possible to support a perspective which entails ontological differences as signs of theoretical innovations. To better specify, if diseases have been *deconstructed* and *reshaped* in the light of stratification, this does not mean that we have simply to extend the taxonomy but rather that such a picture calls for a new taxonomic perspective. In other words, the multilayer situation (genetic and epigenetic background, physiological mechanisms, social and economic factors, life style, etc.) that medical doctors should face in dealing with stratification within the clinical context suggests that many understandings about diagnoses and therapeutic strategies highly depend on the way pathologies are classified. Epistemology arises precisely at this stage. Whereas a new approach, as stratified medicine, brings such innovations that directly affect the foundational categories of the discipline itself, a robust epistemological analysis appears to be crucial.

Classification does not mean just dissecting the world into its parts. Far from being a mere accumulation of data, a classification is neither a neutral nor an absolute operation.[19] It rather involves particular research interests, specific aims and scopes and also values. The way scientists collect and order data is shaping their research since it is shaping the type of questions they pose and how they pose them. This aspect might be unfamiliar to medical contexts, however it is well known in philosophy of science as the *theory-ladenness of observation*.[20] This expression means that scientific observations always 'carry' a theoretical weight. Again the particular order researchers adopt to set up their data is crucial to determine the further analysis of those data and their possible manipulation. As famously claimed by Poincaré '[t]he scientist must set in order, science is built up with facts, as a house is with stones. But a collection of facts is no more a science than a heap of stones is a house'.[21] In other words, classification, as well as observation, is an activity that selects certain particular features of the phenomena under investigation, in order to construct a set (or a map) of specific relationships among these phenomena. For example, in a phylogenetic tree analysis, scientists group species and organisms according to a specific theoretical background, i.e. on the basis of the evolutionary relationships putatively occurring among them. Because of this it should be clear that, beyond the reasons that motivated such analyses, a phylogenetic tree is not the ultimate nor fundamental way to group those natural objects but rather a way of classifying biological entities according to the specific aims of a branch of the life sciences: evolutionary biology.

Coming back to medicine, the way data on pathological conditions coming from *omics* are structured and produced (an operation that involves also which computational and experimental settings researchers want to adopt), shapes the categories on which pathologies are defined. In this sense it is possible to understand the shift from 'leukaemia' as referring to a particular disease to 'leukaemia' as a common, general label which stands for a spectrum of different conditions. Rather from being metaphysically conceived, these differences pertain to different levels of abstraction. 'Leukaemia' is still a viable term to define the whole big category when physicians have to distinguish it from, for instance, solid tumours. Then such a term undergoes an epistemic decomposition due to novel ways of analysing data. This is because from a level of description that was mainly morphological and organ-specific, the classification strategy moved to another one based on *omics*, tissue/cell situated and predominantly molecular. It is not the pathology that is changing as it were a natural kind in the world but it is rather the epistemic strategy that determines which are the ontological categories.

One consequence is that the question about 'what a disease is' cannot be simply posed *per se*. Indeed the answer will depend on the level or granularity of the analysis. Different contexts adopt different levels of granularity. Thus *the problem here is to bridge the gap among these levels*. Bridging the gap here is different from reduction. This is because the aim of such an effort is to formulate hypotheses and elaborate inferences maintaining the different levels of granularity, since, for instance, what is of interest of a clinician should be preserved and not conflated with the interests of the bench-researcher. Thus bridging the gap means also providing translational tools. The problem lies in the fact that different levels of abstraction are difficult to translate horizontally from a context to another. For instance, the translation of such a fine tuned classification of diseases through *omics* technology to the clinical context is not straightforward. Representing and translating knowledge from one level of description to another (i.e. 'from the bench to the bedside' to use an expression that has become quite common) is a key feature in order to address the new challenges of personalized medicine. Let us try to explain why it is so.

Let us consider the example of a disease as cancer. 'Cancer' as a term labels a great variety of heterogeneous phenomena grouped on the basis that they share some common features as, to mention one, the uncontrolled proliferation of certain cells due to the accumulation of potentially dangerous genetic mutations or genomic rearrangements (which will endanger the survival of the organism). At a lower/molecular level of investigation, the manifestation of the disease may result even within the cellular context, since uncontrolled proliferation may be observed already in a single cell. Indeed, for the vast majority of molecular biologists, the cell is the level of explanation.[22] According to this perspective, the disease is already present from the moment that at least one cell starts to behave abnormally. Then the causal chain proceeds from the cellular level up to tissues and organs. At this stage one person could not ever suffer from cancer as the cell might be destroyed by the immune system or simply it could not proliferate. Thus, the mere presence of abnormal cells does not constitute, *per se*, an indication that a person will suffer from cancer. On the other hand, in a clinical context, the disease is definitely a more complex entity which arises more clearly when it affects the functions of an organism at a more systemic level.

Accordingly, the gap here lies in the fact that the data and information produced in one context, namely the experimental setting, cannot be properly treated in the clinical one. Without a proper *integration* it would be impossible to take advantage of the results produced in the lab in order to elaborate precise therapeutic strategies. In other words, scientists and clinicians are facing the huge problem of how it is possible to manage and process together data produced, stored and organized in different manners and often belonging to different data-sets. This aspect explains why so many studies have arisen in order to answer to the need for *integration* (concerning both data and information). However, there is neither a unique definition of integration yet nor a unique form of integration.[23]

Data-integration is generally considered as a technical problem due to the intrinsic differences, within the scientific practice, among the procedures of data production and collection. However, given the heterogeneity of data, especially in biomedical research, not just one approach has been provided. Some scholars suggest that, generally speaking, integration 'encompasses the combination of methods and methodologies [...], the process of making data-sets comparable and re-analysable, and the variety of ways in which explanations are brought together in a particular inquiry'.[24] Moreover, aside from data integration, there are also other questions concerning *methodological integration* (i.e. how to combine, especially in systems biology, different techniques addressing the same problem – the so-called *multidimensional understanding*) and *explanatory integration*, which concerns the unification of different theoretical contributions in order to explain a phenomenon or a set of phenomena and the use of models in a specific field from another area of investigation.

Bio-Ontologies and Integration

In the recent years the problem of integration has stimulated a proliferation of possible solutions to it.[25] An interesting perspective is the one advanced by the implementation of computer science tools called *computational ontologies*. The term 'ontologies' stems of course from philosophy but, recently, it refers also to something very different. While philosophical ontology is just devoted to speculation, engineers and computer scientists revitalized the notion in the light of its possible application. *Applied ontology* is now a promising field of computer science that deals with organizing information. In modern computational jargon 'an ontology is a specification of a conceptualization'.[26] To put it differently, a computational ontology is a way of *modelling* and representing a domain of interest or a particular area of knowledge so that a computer can process it. As Gruber pointed out[27] 'an ontology specifies a vocabulary with which to make assertions, which may be inputs or outputs of knowledge agents (such as a software program)'.

Ontologies are conceived as the mode to annotate and translate a specific knowledge of a certain level of description to other levels through the establishment of a common controlled vocabulary. Therefore ontologies are also said to be the 'semantic level' of scientific modelling. Ontologies can be divided between the so called *upper-level ontologies*, which are concerning the foundational relationships and the basic objects that are applicable to virtually any particular context, and *domain ontologies* which are conceived to model specific area of interest. Domain ontologies capture and map different conceptual contents but they make these contents comparable in virtue of their underlying common structure (which is made explicit by upper ontologies).

Thus *bio-ontologies* constitute a propitious tool in many areas of biomedical research. Bio-ontologies are thought to be a candidate to manage the data diversity as they can, in theory, allow *integration* between different datasets. They can also favour data integration from molecular biology to the clinic by connecting, or better *aligning*, molecular information to pathological and anatomical data and categories. To put it differently, bio-ontologies 'provide a structured, controlled vocabulary through which data – especially those gathered through sequencing and genomics, but increasingly also those resulting from other types of research – can be classified in a form that can be stored in and retrieved from online databases'.[28]

One example is given by *Gene Ontology* (GO), developed by the Gene Ontology Consortium, which represents a promising project in order to provide a representation of the features of gene products across different species and data-bases through a controlled vocabulary of different 'biological categories'. In GO the ontology is structured in three domains: (1) *cellular component* which refers to the parts of a cell or its extracellular environment; (2) *molecular function*, which represents the fundamental activity of a gene product; and (3) *biological process* which represents temporally defined and circumscribed molecular events. Ontologies as GO allow a form of *horizontal integration* because gene products can be represented across different species and data-bases through GO terms which are then localized and contextualized through the practise of annotation.

Another project, *The Open Biomedical Ontologies* developed by a different consortium (OBO Foundry) to provide a common framework among different ontologies (including virtually also GO) within the life sciences in order to provide a sort of *vertical integration*. 'The result is an expanding family of ontologies designed to be interoperable and logically well formed and to incorporate accurate representations of biological reality'.[29] OBO ontologies are all constructed according to the adoption of a common upper ontology which provides a complete interoperability. In other words 'each Foundry ontology forms a graph-theoretic structure, with terms connected by edges representing relations such as '*is_*' or '*part_of*' in assertions'.[30] The relations were consistently developed through a specific devoted ontology called Relation Ontology (RO). Such a common framework allows also *cross-products*, combined terms intersecting different ontologies, which are perfectly consistent and avoiding redundancy.

Beyond the technical details, there is a lot of space here for a genuine philosophical analysis. Again this is a case in which philosophy stems from scientific practise. Indeed although bio-ontologies are called in such a way they are really an *epistemic tool*. This tool highlights the ontological structure of integrated data according to a particular epistemic setting. When Barry Smith rightfully claims that 'ontologies are a window on reality'[31] we have to carefully reflect on what reality is.

Bio-ontologies are a way to represent biological knowledge. Because of that an ontological determination will always rely on a predetermined epistemological framework. Thus these ontological specifications should be thought as a way to make explicit what is there according to the current domain-knowledge.

The question is not trivial. The categories of a bio-ontology do not simply refer to 'objects in the world' but rather represent a high degree of theoretical idealization and epistemic abstraction as they are the result of a complex interaction between experimental findings, technical constraints and semantic analysis. Therefore when researchers refer to Gene Ontology categories they are not simply operating on what philosophers call 'natural kinds'. Gene Ontology categories, as they are representing the integration of experimental data, should not be conceived as 'real entities' but rather as *orienteering tools* and by means of which to elaborate new experimental strategies. As in the orienteering game, bio-ontologies are those tool helping the researchers to find directions in an unfamiliar territory. Bio-ontologies thus provide a map for making the conceptual content comparable across different contexts. This is because bio-ontologies are a form of scientific representation that differs from traditional theoretical models because, since they stand for data-integrations, they also provide a measurement of something. Following van Fraassen 'scientific representation is not exhausted by a study of the role of theory or theoretical models. To complete our understanding of scientific representation we must equally approach measurement, its instrumental character and its role. I will argue that measuring, just as well as theorizing, is representing'.[32] Therefore the interoperability of different ontologies at different granularity does not produce just an integration (as a technological issue) of experimental data to be used in a therapeutic setting. Bio-ontologies are then candidates to allow also genuine epistemic integration since they promote comparison on the epistemic perspectives from which those data have been produced and interpreted.

Conclusions

In the past, many efforts have been made to draw out how much metaphysics is in science.[33] And now it is time to think again the Popperian thesis according to which the philosophical problems have scientific roots.[34] Considering this last point, probably not all the genuine philosophical problems stem from science. Nevertheless, surely, if we analyse science carefully and with philosophically trained eyes, we find in it a lot of metaphysical assumptions and a lot of philosophical implications. In particular, if we analyse contemporary biomedicine through those lens, we immediately realize how much it is embedded in epistemological questions and how much it raises ontological questions, especially as a consequences of the powerful technological tools it can use.

Here above we have focused on the molecular stratification of medicine. We have started clarifying what it involves at the level of drug response and risk assessment. Thus, we have understood the relevance of the classification: the real core of stratification medicine. But classification means, on the one hand, the epistemological question concerning how to classify and, on the other hand, the ontological questions regarding the epistemic value of the classification and the way in which different classifications can be fruitfully correlated for the physician (and thus, extremely important, for the patients). In so doing, we have seen how the old but still living theme of the theory-ladenness of the observation is the conceptual core of the entire matter. If we used Cassirer jargon, molecular classifications should not be intended *sub specie aeternitatis* but *sub specie postulate*.[35]

Summing up, and if a conclusive thought should be made, also in this case we have a strong feeling that a philosophical *logos* detached, or elaborated independently from, the lab *tèchne* and the molecular *bios* cannot be able to meet the real challenges of our time and, therefore, to successfully tackle them. On the other hand, we have also the strong feeling that a lab *tèchne* and a molecular *bios* without the right philosophical (in our case, epistemological and ontological) insight could be just a product oblivious of a long intellectual history, which had its birth in the ancient Greek culture.

7 EPISTEMOLOGY OF ROBOTICS: AN OUTLINE

Giampaolo Ghilardi

Robo-Ethics: Towards an Epistemology of Robotics

To get started there is a paradox to solve with robo-ethics. This concept indeed is composed by two different terms in contradiction one another: robot and ethics.

On the one hand, robot means slave. This word has been used in English language since 1923, deriving from *Rossum's Universal Robots* (1920), a play by Karel Čapek,[1] where for the first time the term made its appearance from Czech *robota,* which means work; the root of this word is related to old Slavonic *robota:* servitude. So robot arises as concept endowed with the ideas of heavy work and slavery.

On the other hand, ethics implies a free agent as precondition of its epistemological status. Quoting the Britannica encyclopedia: 'ethics, also called moral philosophy, is the discipline concerned with what is morally good and bad, right and wrong. The term is also applied to any system or theory of moral values or principles'.[2] In order to be able to judge what is good and what is bad, to discriminate right and wrong, and acting accordingly to these values, as we said, freedom is a necessary requirement.

So the paradox could be expressed in this way: there cannot be an ethics concerned with robots, since robots are not free agents by definition, while ethics requires free agents to develop its analysis.

Nevertheless today we are dealing with robo-ethics as an established discipline with a proper epistemological status. In this chapter we want to solve this issue, by showing under which conditions it is possible to talk properly about robo-ethics.

There are three strategies to overcome the paradox we have just seen: to deny robots are what they are; to deny ethics is what it is; to understand what robot and ethics really are. All these strategies have been followed at some degree. From a mere logical point of view a successful attempt to solve the question has to widen the boundaries of robotics and ethics to make room to a merge of both these fields, and this operation on one hand could look like a denegation of these disciplines, while on the other hand it should be accomplished by a deeper understanding of both robotics and ethics.

Let us show some examples of what we are talking about.

Moral Machines

Wendell Wallach and Colin Allen in their book *Moral Machines: Teaching Robots Right from Wrong*[3] try to demonstrate the possibility of future Artificial Moral Agents (AMAs):

> Wallach proposes a continuum of moral agency for all technology, from everyday objects completely lacking agency to full-fledged sentient robots with full moral agency. The continuum exists along two dimensions: autonomy, which indicates what the technology has power to do, and ethical sensitivity, which reflects what inputs the technology can use to make decisions. For example, a hammer has no autonomy and no sensitivity, while a thermostat has sensitivity to temperature and autonomy to turn on a furnace or a fan.[4]

It could be debatable whether one could claim autonomy for a thermostat, since it lacks a proper decision on different options; it is true it performs a choice, but it is mechanical one, a deterministic one, that's why we cannot acknowledge its operating system as autonomous.

Broadly speaking, Wallach and Allen choose an approach that allows the artificial intelligence (AI) to model human moral development. They seem to take seriously evolutionary psychology.

> The system attempts to learn appropriate responses to moral considerations in bottom up approaches, which take their inspiration from evolutionary and developmental psychology as well as game theory. Instead of selecting a specific moral framework, the objective is to provide an environment in which appropriately moral behavior is developed, which is roughly analogous to how most humans 'learn' morality; growing children gain a sense of what is right and wrong based on social context and experiences. Similarly, bottom-up approaches, including evolutionary algorithms, machine learning techniques, or direct manipulation in order to optimize a particular outcome, can be applied to facilitate a machine achieving a goal.[5]

It is worth noting that from the perspectives of both evolution and individual psychology, the question of how human beings become moral is not uncontroversial. But, at the very least, it seems to be an *empirical* question, with the available theories more conducive to being programmed into a machine than moral theories like virtue ethics, utilitarianism or Kantian deontology.[6]

As we can see, the main argument of this approach is the supposed learning capability of machines which could be taught to be moral by experience. It was Norbert Wiener, the father of cybernetics, who said in his *God and Golem* that: 'There are at least three points in Cybernetics which appear to me to be relevant to religious issues. One of these concerns machines which learn; one concerns machines which reproduce themselves; and one, the coordination between machines and man.'[7]

There is a strong emphasis nowadays on the so called bottom-up approach, which is considered to be more empirically based than the top-down one, which is labelled as rationalistic.[8] The fact that robots can learn some behaviour patterns needs to be understood in order to see if this operation could really be considered as learning. When we assume the learning metaphor in association with a machine activity we are, whether aware of it or not, anthropomorphizing our products. Anthropomorphizing is not necessarily a mistake or something one should to avoid in every circumstance, rather it is an operation that requires particular care. In this case, the machine does not really learn anything, it rather stores data which will be used again whenever the environment will present the same features. Of course data recording is a part of the learning process, and the same goes for applying the gathered schemes in a given environment, but these steps do not exhaust learning, which implies also the skill to creatively adopt different schemes in similar situations. In other words, the central point of learning is the analogical reasoning, which is not some skill, rather it is the very essence of intelligence. We are not now arguing about the real possibility for an AI to achieve such a kind of intelligence in future, we limit ourselves to note that it has not yet been achieved, and for this reason we need to pay attention to understand what we are talking about when we adopt the learning metaphor.

Awareness of this should make it easy to see that morality cannot be learned by a machine in proper terms; machines could at their best store some behaviour pattern that could be applied at a given time, and thus reply an apparent moral outcome. This appearance of morality is interesting to analyse, as it brings to mind Alan Turing's[9] famous experiment exposed in 'Computing Machinery and Intelligence'. It is worth noting that such a theory came from considering the externally observable behaviour of a human computer, a person who carried out computations with pen and paper, and 'is supposed to be following fixed rules', so that Turing modelled what a person does, not what a person thinks. This is exactly what happens in a robotics context when we talk about machines' morality. Since the robots do what they do not for the sake of an inner intention, but for an algorithm which has been programmed and installed in their CPU, we can, at our best, talk about their behaviour as something that externally resembles a moral behaviour, just as in Turing's experiment, where we cannot distinguish at first sight who or what is carrying out computations.

Is resembling a moral behaviour sufficient reason to be defined as moral? Maybe from a functionalist point of view it is, but, then again, is the functionalism[10] a satisfying perspective in ethics? We think it is not, for the above mentioned reasons: ethics cannot avoid to consider internal constitution of mental states, intentions are the true objects of moral arguments and cannot be simply neglected just because they are not identifiable as behaviour.

Slave Morality

Another interesting attempt to overcome our issue is the one adopted by the so-called slave morality approach. It could be summarized as such: they [robots] have no ends of their own, but their goals are all in service of the goals of someone else, the military or, more specifically, whoever commands them and gives their orders. Robots cannot create their own laws or final goals; they are not ultimately makers of a *self*, but followers of 'life goals' others have imposed, and their own freedom only comes in the mere means they choose to realize those ends.[11]

As we can see here it is pointed out in precise terms the kind of freedom a machine is allowed to perform, and, in Kantian terms, we could say it is about means, not about ends. A robot can act according to someone else's goals, and this means he is following a hypothetical imperative, not a categorical one, the only one that would have allowed us to consider its morality.

We can explore this concept in more detail by observing the difference with man:

> There is a crucial difference between a human soldier or even a human slave and a robot programmed with a slave morality: the human person, whether a soldier or a slave, is presumed to have the ability to disobey orders, even if the punishment for doing so would be harsh. But a properly programmed robot with a 'slave morality' literally could not disobey orders intentionally.[12]

Disobedience is a concept that can make sense only in the context of a categorical imperative, to stick with Kantian vocabulary, since to infringe a given rule in order to achieve a fixed goal is more a matter of miscalculation than something of moral relevance.

As far as we can see this perspective is not a real solution to our epistemological problem about robo-ethics, rather it is useful to point out the limits and the capabilities of deploying machines in moral context: maybe it's easier to consider this condition under the category of 'instrumental morality', which implies a certain degree of freedom about the instruments it is possible to choose, rather than about the goals.[13]

Of course ethics in this sense is not adopted in its full meaning, it is rather considered in a very peculiar way akin to the utilitarian perspective.

This point of view is useful because shows us how freedom could be involved in a certain degree without a proper free agent, and this trace will constitute the path leading to the solution to our starting paradox. In order to achieve this goal, we are adopting an analogical meaning of freedom, in the same way it is used in mechanics when it comes to talk about Degrees of Freedom (DOF): 'The number of independent inputs required to determine the position of all links of the mechanism with respect to ground'.[14]

It is clear that freedom is not a number, nor it can be described in numerical way, still the analogical adoption of this concept is consistent with the idea of freedom itself. In this case we see a cinematic system with different movement directions, and this variety provides the conditions under which it is possible to talk about freedom. A ball falling down along a narrow tunnel is not free to choose its path, nor is it free to choose whether it is falling or not. Whether it is falling or not is not up to the ball to decide, and in our previous terms this could be considered the categorical aspect of the dynamics: once the first move has been made, the directions become an object of 'choice', and at this level the more options are available, the higher the degree of freedom. This is not a full freedom, of course, but neither is it something too far from its very concept, since it is true that the more choices are available, generally speaking, the more free is the context. Even if it is true that freedom does not limit itself to a selection of choices, but implies a wider range of meanings, nonetheless this specific meaning cooperates to describe its nature. This is the reason why this analogical use of the concept, though partial, is not improper.

So, back to the topic, slave morality, which is a bit contradictory in itself, though it can be rescued from its contradiction by referring to an analogical meaning of instrumental morality, where the degrees of freedom allowed to the robots are not about acting or not acting, but rather about how to carry out predetermined ends.

Let us add a consideration about instrumental morality. We are aware this kind of vocabulary is strongly imbued with an utilitarian perspective, which we do not consider to be theoretically consistent. By naming instrumental morality we simply refer to that sphere of morality concerned with the means rather than the ends of a given action, since we assume that both means and ends constitute the object of moral philosophy, according to the Aristotelian ethics.[15]

Asimov's Laws

Asimov's laws[16] have been investigated and analysed by a great number of authors, not to mention that Asimov himself wrote more than eighty short stories exploring how many unexpected and potentially dangerous conditions arise from the combination of these rules.[17] We do not want increase the number of

Asimov's commentators, we would rather consider his work in the light of our paradoxically starting point: how ethics can possibly raise from lack of freedom.

In this respect Asimov's scenarios are an interesting object to study; we can see his stories operating a sort of Kantian answer to the liberty question whereas the ultimate robotic laws are not somehow just the boundaries of the robot actions, rather, they positively indicate what a machine can do, drawing its action range and for this reason setting its liberty space.

Interestingly, Asimov's arrived at his three laws from his *positive* view of the role played by robots in society:

> In the 1920s science fiction was becoming a popular art form for the first time ... and one of the stock plots ... was that of the invention of a robot. Under the influence of the well-known deeds and ultimate fate of Frankenstein and Rossum, there seemed only one change to be rung on this plot – robots were created and destroyed their creator ... I quickly grew tired of this dull hundred-times-told tale ... Knowledge has its dangers, yes, but is the response to be a retreat from knowledge? I began in 1940, to write robot stories of my own – but robot stories of a new variety. My robots were machines designed by engineers, not pseudo-men created by blasphemers.[18]

So, according to the author, the famous three robotic laws have been invented to save science fiction from its apocalyptical origin and to be placed on a more scientific and optimistic ground.[19] On the one hand, freedom is defined as independence from the 'compulsory will of another'; on the other hand, this independence need to be softened in order to allow the coexistence of all, and this is possible by the sake of a universal law which is stated in these terms: 'act externally in such a manner that the free exercise of thy will may be able to coexist with the freedom of all others, according to a universal law'.[20] We won't venture any further into Kantian philosophy, which has its own apories. What we want to point out is the solution which has been given to problem of liberty grounded on the domain of universal law. This little window opened on Kant and Asimov's work help us to cast a bit of light on our topic. On the one hand, according to Kant, if freedom has to be conceived as 'independence from the compulsory will of another', then a robot by its own definition cannot be considered as a free agent; on the other hand, if this independence is achievable following an universal law, then we can see a viable way of escape, the one adopted by Asimov. Of course the theoretical cost of such a solution is a sort of formalism, which is inescapable in the Kantian philosophy, and which has been literary explored in Asimov's stories where all the contradictions and paradoxes emerging from this kind of approach have been shaped in a new kind of literature. The deep question that lies at the root of this philosophy is: how can a law, universal or not, set someone or somewhat free? The question arises because it is difficult to see a law as something internal to the one who needs to conform to it; as long as a law is not internal to those who have to follow it, it cannot be considered as something which can set anyone free. These sort of issues spring from a rationalistic mindset that divides form and matter too harshly. The very idea to set laws in a software which will command the hardware stems from a Cartesian perspective, where we have a *res cogitans* (software) on one side and a *res extensa* (hardware) on the other side. This sort of division, which has revealed itself as a cultural disaster in many respects, does not reflect reality as it is and neither as it works, since none of us have ever experienced a disembodied mind, or an in-animated body. It is always the rationalistic erroneous mindset which considers liberty as something essentially related to the realm of mind, and this is why the Kantian Asimov's approach is laws-based. Laws in robotics play the role of the mind in ethics. Considering freedom as something essentially related to mind and not committed to the body is a naïve mistake, as well as an unempirical evidence, since the first experience of freedom is related to the body. We can see another echo of this dualistic perspective in the already quoted Kantian sentence: 'act externally in such a manner that ... '; the stress on the external acting recalls to the mind the Turing experiment and its request to consider the behaviour, not the inner reasoning. These remarks share a common dualistic ground that affects every conclusion and which does not allow us to reconcile the world of matter with the reign of mind, and the field of robots with that of humans. In his way, Asimov, although still holding the rationalistic bias that has been mentioned, did not limit himself to present a top-down ethics model, but instead showed us the unavoidable arising conflicts connected to that paradigm. As we saw through Asimov's very words, his intention was to skip the ascientific and somehow religious embedded point of view of his former colleagues such as Shelley or Čapek, and he tried to overcome that perspective by the means of science. A science that, in the 1950s, was still dominated by Cartesian biases. In this respect the new bottom-up approach, recently developed in robotics under the definition of embodied intelligence, can shed a new light on these issues.

Value Principle

Strictly speaking, the value principle is not a robo-ethics theme, despite its appearance: although both the concepts of value and principle are fundamental in ethics, they are adopted in robotics without a specific moral connotation. They are, rather, used as intellectual tools to understand and design the properties and principles of intelligent systems, and this is also the reason why we are interested in them. We have noted how design involves many assumptions that are both epistemologically and ethically sensitive.

The value principle states that intelligent agents are equipped with a value system which constitutes a basic set of assumptions about what is good for the agent.[21] Interestingly, one of the first remarks made about the value principle in Pfeifer's work is that:

> The designer decides that it is good for the agent to have a certain kind of locomotion (e.g. wheels), certain sensors (e.g. IR sensors), certain reflexes (e.g. turn away from objects), certain learning mechanisms (e.g. selectionist learning), etc. These values are implicit. They are not represented explicitly in the system.[22]

This quotation points out that what is good for an 'agent' has not been decided by the agent itself, rather it has been pre-set by its designer. In a Kantian vocabulary this framework would have been described as an hypothetical imperative, the sort of command specific to non-autonomous agents. To make this point clearer, consider what being autonomous means: it is not just a matter of self-government, as usually stated, it is, rather, something related to being in charge of the choice of proper own goals. According to the etymology of autonomy, to be autonomous (αυτόνομος which comes from αύτο, *auto,* 'self' and νόμος, *nomos,* 'law') entails being a law unto yourself, which is exactly the opposite of machines which move themselves following the designer's pre-set goals.

According to the 'value' principle, the learning mechanisms have to be based on principles of self-organization, since the categories to be formed are not known to the agent beforehand.[23] This quotation about self-organization is a key concept to understand the logic of value principle. What needs to be highlighted is that the goals around which self-organization takes place is the one given by the designer.

We do not here need to discuss the paradigm shift that brought perception to be thought of as related to sensory-motor coordination rather than a process of mapping a proximal (sensory) stimulus onto some kind of internal representation, even if this was the conceptual trail that marked the beginning of embodied intelligence. It is enough for our purpose to elucidate how this new way of designing robots, while on the one hand aiming to develop more autonomous machines by the means of the value principle, on the other hand never ceases to recognize that 'This interpretation in terms of value is only in the eye of the designer – the agent will simply execute the reflexes.'[24] Authors who suggest that we pay attention to value principles are aware of the role played by learning in this approach, and they try to answer the questions 'How is learning related to value? Why continue to acquire more and more sophisticated skills and not be happy with what you have?',[25] referring to the motivated complexity principle, where motivation, according to their words '[s]hould emerge from the developmental or evolutionary process'.[26]

Let's see how this paradigm tackles the learning issue related to motivation aiming to drive some conclusive remarks. Dealing with the idea of flow that was developed by Mihalyi Czikszentmihalyi,[27] it is noted that

> By analogy ... If the organism masters one skill, its processing demands will decrease
> and it is then free to use exploration strategies to increase its inflow of information
> ... In other words, the more skills the organism has mastered, the more readily it can
> indulge in exploratory activities.[28]

Here we can see how the learning process, besides providing the 'self-reference' necessary to action, sets the agent on a higher level of freedom, allowing it to explore and engage with other tasks to be learned, in a virtual infinite learning movement.

A key feature of the value principle is its vagueness,[29] as stated by its coiners, but this characteristic, far from being a bias, or worse, is an intentional open window meant to allow the machine to explore as many ways as possible to accomplish its task rather than limiting its capabilities to an established set of choices that are preplanned, once and for all.

We should also add that, as much as this principle could be appealing, it has not really been used in any machine project. As far as we know, no robot is actually working on a value principle project that can be considered as an inspiring way of reasoning, but still not an established projecting practice. Of course, this principle cannot fit a top-down approach; it cannot be written in an algorithmic syntax, since a principle, due to its very nature, cannot be translated in a one-to-one relation. The end will always transcend the means by which it can be achieved, this is the reason why it cannot be represented in an univocal way.

In this respect the end can only ever be in the eye of the designer, as stated by Pfeifer, not in the software running the machine. The practical problem to design a system aiming to pursue an end not limited to a set of representations, cannot be faced by means of a classical AI approach. It should rather be addressed through a bottom-up approach, such as the one deployed in an EI perspective, considering that the end will never be explicitly encoded, and therefore it cannot be recognized as effectively achieved once and for all.

This aspect of the value principle closely resembles a well-known ethical virtue: the practical wisdom, also known as *Phronesis*.[30]

Let us set about making this aspect clear:

> Aristotle makes it clear that phronesis is not the same as *theoria* – that is, a theoretical
> knowledge that is propositional and learnable in a purely intellectual way. Phronesis
> cannot be programmed into a computer ... It is, rather a kind of 'know-how'. It is not
> reducible to a set of rules, however, and should also be distinguished from *techne*. The
> good person, the person with phronesis, sees what to do in an immediate way, and
> does the good thing in a close to automatic way, as if it were second nature.[31]

Aristotle's analysis on *phronesis* is an interesting chapter of his *Nicomachean Ethics*, where the philosopher asks himself how it is possible to learn *phronesis*. He answers the question by saying that one could learn *phronesis* by hanging around with good people. What is important to Aristotle is making clear that such a virtue is not gained as normal knowledge, but can be learned just by acting. This statement, however, leads to an apparent contradiction: how is it possible to act in a good way if we do not know what good is? The problem was solved by Aristotle distinguishing good in general things from good in particular cases, and considering how phronesis is a virtue connected with good in given situations, the capability to take the right choice among others in order to achieve good for the agent. In a certain sense, in value principles we find the same challenge: as a principle our set of values is not a matter of choice, neither it is possible to write it in a calculabilistic way, as a practice it could be deployed as a computation[32] among the different options the machine has to face from time to time.

Analogy

It has to be clear that expressions such as *moral machines, robo-ethics, techno-ethics, intelligent robots* and *embodied intelligence* are used in an analogical manner, where is not just a rhetorical figure, but, rather, a logical one.

The complex theory of analogy tells us that we need to know which is the '*princeps analogatum*' in order to correctly understand the analogies. In this case it is the man, since we can see applying specifically human categories to not-human objects. Both ethics and intelligence are essentially human predicates, when these attributes are going to be used to define something different than men, then we need to pay attention how this switch is accomplished and what consequences will bring with.

We do not want to give the misleading idea this is just a vocabulary problem, even if there is an urgent need to clarify terms. We think things could be easier considering a few points related to the idea of intelligence. Intelligence, according to its Latin etymology (*intus legere*), is basically the power to read something from within;[33] it is a kind of view of the internal structure of the object we are looking at. So intelligence could be considered as a sort of faculty to grasp the hidden feature of reality, the nature of what stands in front of us. Bearing this in mind, we can see to what extent the analogies which are intelligence-based have to be understood to not be misleaded by their adoption. Let us suggest some examples. Talking about artificial intelligence is always challenging and fascinating because both the elements of this concept are somehow opposite on another: on the one hand, it is an artefact, since it has been made by some other agent, does not have in itself a proper intention, and at its best can have a pre-set programme; on the other hand, intelligence by its own definition is an intentional act that proceeds from the inside out, while an artefact not only has been assembled from the outside, but proceeds in the opposite direction: from the outside in.

Another example is the already mentioned embodied intelligence. In this case the analogy between body and intelligence is more grounded, because even the body shows a movement from within; what get lost in this image is the act of reading characteristic of intelligence. In a certain respect this aspect does not get entirely lost; indeed embodied intelligence could be considered a particular way to read the environment in order to react and to adapt to it. Although this kind of reading is not strictly speaking the one meant in the very concept of intelligence, nonetheless it is always a possible meaning of it, hence the analogy is legitimate.

As we have seen all these analogies are grounded in the anthropological domain. The robo-ethics epistemological issue we started with can be easily understood if we bear in mind these few remarks: there are different kinds of robotics, each one encompassing a particular concept of intelligence; there are different kinds of ethics, each one with different concepts of good and evil, right and wrong, etc. Every kind of variation on these subjects entails changes to robo-ethics which is going to be deployed.

As we stated at the beginning of this chapter there are different strategies adopted to overcome the robo-ethics paradox. Those who talk about moral agents include robots in this category, and on the one hand aim to extend the domain of morality beyond the classical boundaries of the human beings, referring to a more neutral 'agent'; this is an abstraction and therefore turns itself into a general concept able to be applied to other entities and not just to human beings. On the other hand, they reduce the dimension of acting to the one of operating, which is more suitable to robots.

Although we understand the reasons underlying this approach, it seems that concepts like agent and morality have been stretched a bit too much, turning out to mean something basically different from their origins. Action indeed is something essentially related to the man who, acting, affects both himself and the object of his action. In classical vocabulary this duplicity of action was expressed in terms like *actio immanens* and *actio transiens*. The former is defined as an activity, an action coming from a given subject and which remains in it, without any influence from the outside.[34] Thus, both the 'source' or 'principle' (*principium*) of action, and the '*terminus*', meaning the result of the said action, are to be found in the subject.[35] On the contrary we have the *actio transiens*, whose *terminus* (result) is to be found outside of the operating subject. In other words '*the object* [of an activity] *is found outside the active subject itself*'.[36]

A good theory of action has to take into account both these aspects of the acting. If it does not, it fails to address its goal, which is to describe what an action is and how can be carried on.

It is worth noting why it has been possible at a certain degree to misunderstand the nature of the machine operational behaviour with action. As we said, action implies a sort of self-reference, the effect of the action falls upon the agent, other than upon the object (that incidentally could be the subject himself, and in that case we will have just an *immanent* action, without the *intransient* aspect). Where do we find this self-reference in machine behaviour? In learning. It has already been spoken of, and it has also already been mentioned how the metaphor of learning needs to be carefully taken into account. Here we want to point out how this metaphor is essential to determine the robot operation as an action, since through learning it seems its operations fall back upon it, increasing and therefore changing from within its structure. But, as we have noted above, the learning we are talking about when it comes to machines is not a real learning, it does not imply a new way to consider problems and solve them. It lacks a real creativity, a real intelligence of the issues, it is all about data gathering and scheme application. Of course, the more data you can get the better you can apply a given scheme, or even change the scheme, but this has nothing to do with learning, since you cannot create a completely new 'scheme', and most of all the process of creating a new scheme, besides being far beyond the capabilities of actual machines, is deeply rooted in that primary meaning of reading we previously mentioned: we need a true understanding of reality in order to conceive an operational scheme that can deal with it. Without this dispassionate search for reality, without this curiosity about what we are facing, we simply cannot have a true knowledge about anything, and therefore we can't develop any scheme, model, representation or theory.

For these types of reasons we cannot accept the model of moral agents applied to robots, not only because robots as such cannot be moral, rather than they cannot be agents. The very concept of agency is not compatible with that of the machine, and, as correctly stated by the authors who coined this idea, we cannot have a moral behaviour without proper agency.[37]

We are not arguing against the use of analogies in this field; on the contrary, we refer to it as an essential working tool to deal with these concepts. Its nature, however, has to be correctly understood, and this means seeing to what extent we can apply specific anthropological characteristics to artefacts. This also means coming to a better understanding of what an artefact truly is, what kind of autonomy it can achieve and what kind of operations it can perform.

Robotics

As we have seen, robo-ethics could be considered both as the ethics of robots and as the ethics of robotics. As discussed, we find the latter definition more suitable than the former. This leads us to take a closer look at what robo-ethics is, and the ways in which it is compatible with ethics.

According to the IEEE definition:

> Robotics focuses on systems incorporating sensors and actuators that operate autonomously or semi-autonomously in cooperation with humans. Robotics research emphasizes intelligence and adaptability to cope with unstructured environments. Automation research emphasizes efficiency, productivity, quality, and reliability, focusing on systems that operate autonomously, often in structured environments over extended periods, and on the explicit structuring of such environments.

As we can see, the main difference between robotics and automation research lies in the kind of environments with which they have to cope. While automation operates in structured environments, robotics has to deal with unstructured environments. This remark helps us to point out a fundamental and characterizing element of this discipline: indeterminism. Unpredictability of the environment is a central point in the robotics field. The main aim of robotics could be summarized as 'striving to improve the machine performance in the face of the unknown'.

It is important to make clear that the indeterminism we are talking about is not just a temporary situation, something that could be overcome once we have gathered enough information to make the environment more predictable. It is, rather, a permanent condition under which the machine has to operate. Unstructured environments are not not-yet-structured environments, rather they are never-to-be-structured ones. In philosophical terms we need to understand indeterminism here as not just an epistemological feature of our context, but rather as an ontological dimension of reality which we need to handle. It is not just a lack of knowledge that determines indeterminism, rather it is the reality that in itself cannot be reduced to our parameters.

Indeterminism, then, is something more than a simple situation, which eventually could be turned in a more predictable one. It has to be considered as a sort of principle, and like in the Heisenberg's principle, the human being is a key element to understand it. Man is a source of uncertainty in robotics as well as in quantum physics; he changes environmental coordinates with which the machine has to cope. This is the reason why a good machine design needs to take human variables into account, and moreover needs to model human goals and patterns. Good interaction between machines and human beings could be assessed when both of them share the same model.

This is why bio-mechatronics is a viable and epistemologically coherent robotics paradigm, precisely because of the man in the loop of design.

As we have seen, all the analogies previously sketched are rooted in the human domain. In bio-mechatronics we have all the dimensions involved – the human one, the mechanical one and the cybernetic one – and a proper and independent role has been recognized for each of these domains, and still they are all sharing a common design.

In the top-down bio-mechatronic design approach, we find, first of all, the analysis of the person–activity interface, then the optimal allocation of roles, then the interaction design, then the functional requirements, then the human component model, then the mechatronic machine design and, finally, the design optimization by analysing performance of the integrated human–robot system.[38] These steps highlight the deep interconnection among the different levels taken into account and therefore the unavoidable complexity of the system. Still, complexity does not mean complication, just that every level of the analysis has its own specific logic which is developed according to the ultimate goal of the whole system.

We have outlined this methodology to show an example of robotics that encompasses both the human element and technology in a way that does not confuse these two domains, but still keeps them together.

At least others two different approaches have been developed in order to overcome the indeterminism problem. The first who tried to face it and wholly changed the way we think about the issue, was Rodney Brooks, who left information in the environment, rather than embedding it in machines. He was a pioneer of the bottom-up approach, since he criticized the representational conception of intelligence, aiming for a more practical and suitable one: 'The intelligent system is decomposed into independent and parallel activity producers which all interface directly to the world through perception and action, rather than interface to each other particularly much'.[39] The focus in Brook's idea switches from representational models to operational features, to be performed through perception and action. This was also, of course, the root of what later in years was named embodied intelligence. Even in this perspective the human element is still missing, while on the one hand it has been correctly pointed out that the old fashioned top-down approach was in dire need of an afterthought, on the other hand the intelligence metaphor kept on being used without questioning the *princeps analogatum*: man.

Another interesting approach to facing indeterminism that is connected to mature robotics is probabilistic robotics, which is: '[a] new and growing area in robotics, concerned with perception and control in the face of uncertainty. Building on the field of mathematical statistics, probabilistic robotics endows robots with a new level of robustness in real-world situations.'[40] This perspective on the one hand has the merit of considering the field of indeterminacy, on the other hand mistakes the level analysis standardizing non-human sensorial inputs with human ones. Not recognizing the human level as a distinct and hierarchically superior one is a design bias which needs to be pointed out.

The bio-mechatronic paradigm previously outlined develops a top-down methodology where the Bio element is embedded into design level, not into the sensory-perception area.

Let us say that an intermediate line could be drawn between automatic systems and autonomous ones, where we find the man in the loop expressly planned at design level.

An example of this dialogue man-machine could be found in domotics: while we can leave it up to the machine to tidy the kitchen entirely by itself, it would be better if it were left up to the person to choose what to eat.

Conclusions

At the end of this journey through some of the current approaches to robo-ethics, we would like to draw some conclusions based on the analogical reasoning discussed above.

First of all,[41] we experience the need to distinguish at least four levels of reasoning which need to be taken into account to develop an adequate robo-ethics: the logical one, the ethical one, the epistemological one and the educational one.

Each one of these different levels has to be considered as being interdependent with the others, so that what we can achieve in the logical field can be applied either in ethical and epistemological domains. The same goes for epistemology, whose findings unavoidably self-reflect on the other disciplines involved.

In logic we have seen the most coherent approach to be the wise and conscious use of analogy, bearing in mind that the adoption of almost every key concept deployed in robo-ethics has to be carefully weighted, since its origin is not within robo-ethics. Most of all, analogy is mandatory to understand how to develop an ethical reflection in an un-ethical context, without disfigure both ethics and robotics.

On the specific ethical field we can say that the moral agent question should be regarded as an anthropological issue, since the only remaining real moral agent is man, and not the robot. Besides, borrowing the bottom-up approach from epistemology is suitable for a new Aristotelian understanding of practical philosophy, where the ethical choice is not considered as a simple match with given standards, but rather a computation among different goal-oriented options liable to get better every time.

Different epistemological paradigms, as we have seen, play an important role in this discussion. The historical artificial intelligence debate shaped a way of conceiving intelligence and therefore ethics, and, more important for our topic, a way to conceive the embodying of ethics in machines. We studied some aspects of the embodied intelligence emerging paradigm, trying to grasp the differences this model implies for ethics. We drew a parallel between a computational model and a calculistic one, noting how the former could be deployed to virtually endlessly improve the machine operationality.

This being said, we need to add that a huge set of competences are needed to govern robo-ethics. Above all, issues arising in this field cannot be faced through a simple training course in applied ethics, since it is not only ethical questions that are at stake, but also logical and epistemological ones. Besides, as we have seen, learning is a crucial theme in robotics, so that it needs to be taken into account even in the educational field of designers, and this will be an agenda for the next work. Moral engineers for moral machines.

8 PREDICTION AND PRESCRIPTION IN BIOLOGICAL SYSTEMS: THE ROLE OF TECHNOLOGY FOR MEASUREMENT AND TRANSFORMATION

Wenceslao J. Gonzalez[1]

Complexity is one of the relevant features of biological systems, whose structure and dynamics can be grasped through biological sciences as basic sciences. Thus, complexity is a key issue for the possibility and reliability of predictions in these systems of nature. In addition, insofar as prediction is the previous step for prescription in science,[2] complexity turns out to be a crucial factor for the patterns of prescription. In this regard, the guidance to solve concrete problems of biological systems, which is characteristic of applied science, depends on the structural and dynamic complexity of these systems.

Moreover, when complexity in biological systems is disorganized instead of organized, there are additional problems for reliable predictions in biological sciences as basic sciences and, consequently, new difficulties for suitable prescriptions in these disciplines as applied sciences. Furthermore, the development of these tasks in biological sciences – prediction and prescription – requires the role of technology to measure the biological reality available. Besides, technology can have a role in the transformation of the biological reality, which needs scientific knowledge with prescriptions on what should be done for solving specific problems. In these cases, the transformation of the biological reality should take into account values (ethical, social, etc.).

Because the philosophico-methodological analysis of evolutionary biology is already made, including the study of the historical trajectory of evolutionism and its influence on philosophy of science,[3] this chapter deals directly with the topic of prediction and prescription in biological systems. In this sphere, complexity – structural and dynamic – is central for the scientific research in biological sciences, and the dual role of technology – measurement and transformation

– should not be avoided in the analysis. Within these coordinates, the chapter follows four main steps: (1) complexity and biological systems, (2) prediction in the case of biological systems, (3) prescription in dealing with biological systems, and (4) the role of technology for measurement and transformation.

Complexity and Biological Systems

Undoubtedly, a central feature of biological systems is complexity, which is commonly linked to evolution. Moreover, this was the origin of the claim that 'everybody seems to know that complexity increases in evolution',[4] which requires analysis.[5] *Prima facie*, this increase in complexity in biological systems may be twofold. On the one hand, there may be an enlargement of the *structural complexity* in biology, insofar as living beings have now a larger variety of structures than at the beginning of the evolutionary process. On the other, there may be an extension of the *dynamic complexity*, insofar as the evolutionary process (either gradually, such as in Charles Darwin's approach,[6] or by other means, such as the 'punctuated equilibria'[7]) originate novel forms of living beings (such as new 'species'[8]) as well as changes in the relations among the living beings.[9] In principle, the dynamic complexity can produce novelty in the procedures (i.e., new ways to obtain changes) or in the results (i.e., new aspects of a given reality or a new reality as such).

Prominently, complexity is one of the features of biological systems that has received more attention from the scientific perspective.[10] Meanwhile, complexity is a particularly relevant characteristic from the philosophical viewpoint. This can be seen from two main angles of analysis. (a) Complexity can be structural, when the biological traits are associated with the configuration – organized or disorganized – of a biological system as such (i.e., the articulation of a given part of the natural reality) and, therefore, as different from other biological systems. (b) Complexity can be dynamic, when the characteristics of the system evolve over time, which leads to a new aspect of the biological system considered or to a novel system as such.

Both sides of complexity – structural and dynamic – can be studied according to two main philosophical ways: epistemological and ontological. An analysis of the epistemological and the ontological ways of complexity from a general perspective is available in Nicholas Rescher's *Complexity*.[11] His view is directly focused on the structural facet of complexity rather than on the dynamic dimension of complexity. Thus, his analysis requires the additional consideration of the dynamic traits of complexity in order to get the full picture of the problem.[12] Rescher's distinctions are here the starting point for clarifying the issue of complexity in biological systems. This contribution is particularly relevant in order to grasp the philosophico-methodological difficulties for prediction and prescription in the biological sciences.

Modes of Complexity and the Diversity in Biological Systems

Epistemologically, the modes of complexity can be diversified in three epistemic directions: descriptive, generative and computational.[13] (1) *Descriptive complexity* is when the attention goes to the length of the account to be given to provide an adequate description of the system at issue (i.e., a microbiological system requires less description than a macrobiological system). (2) *Generative complexity* is the length of the set of characteristics to be given to provide the system in question (e.g. the attempts of creating artificial life start commonly with the less complex forms of life). (3) *Computational complexity* is the amount of time and effort involved in resolving a problem, which is bigger in the case of climate change than an issue on molluscs in an estuary of the coast of Galicia (north-west of Spain).

Ontologically, there are three main modes of complexity: compositional, directly structural and functional.[14] The first ontological mode – the *compositional complexity* – has two possibilities: constitutional and taxonomic diversity or heterogeneity. (a) Constitutional complexity depends on the number of constituent elements or components (e.g. in its morphology),[15] and makes some research on some biological systems more difficult than other. (b) Taxonomic diversity or heterogeneity emphasizes the variety of constituent elements (i.e., there is a number of different *kinds* of components in their biological configurations), which requires focusing on the diversity inside a biological system.[16]

Within the second ontological mode, the *directly structural complexity*, there are also two factors to be considered: organizational and hierarchical. (i) The organizational complexity is the variety of different possibilities of arranging components in diverse modes of interrelationships (e.g. in the context of survival).[17] (ii) The hierarchical complexity elaborates according to the subordination relationships in the modes of inclusion and subsumption or the organizational disaggregation into subsystems (e.g. molecules, cells, tissues, organs, organisms, populations, communities, ecosystems, etc.).[18] In this regard, 'the basic idea is that higher-level entities are composed of (and only of) lower-level entities, but the prevalent concept of hierarchical organization involves stronger claims as well'.[19]

Concerning the third ontological mode, the *functional complexity*, there are again two possibilities: operational and nomic. (1) Operational complexity looks at the variety of ways of operation or types of functioning (e.g. whales have a more complex lifestyle than barnacles). (2) Nomic complexity is the elaborateness and intricacy of the laws governing the phenomena in question (e.g. large ecosystems are more complex in this manner than small ecosystems). Both possibilities – operational and nomic – have a long tradition in the research of biological systems.

Obviously, when the system develops certain operations or follows some laws, these operations or rules can be more or less complex. Furthermore, the changes over time might be somehow shallow ('evolutionary') or clearly deep ('revolutionary') as well as continuous or discontinuous. In this regard, complexity in dynamic terms is clear when the systems are goal-directed in their *modus operandi*, as is commonly the case in the sciences of design.[20] A feature of complexity is that they do this 'generally towards a plurality of potentially competing goals'.[21] Rescher thinks that there might be two options in the complex system: an *operational* complexity, which is 'displaying dynamic complexity in the temporal unfolding of its processes', or it may be *nomic*, which is 'a timeless complexity in the working interrelationships of its elements'.[22]

Ways of Complexity in Evolutionary Biology

Regarding evolutionary biology, there was wide acceptance that evolution had produced complexity. In this regard, the claim that complexity increases in evolution was studied by Daniel McShea, who considers 'that not enough evidence exists to make an empirical case either for or against increase'.[23] In addition, he makes explicit that 'complexity' and 'progress' are different concepts. Thus, he asks that 'the reader do not equate complexity with progress',[24] although evolutionary trends are frequently related to an increasing adaptability and a growing control by organisms over their environment.[25]

Underneath the idea that complexity increases in evolution seems to be both sorts of complexity: the structural dominion and the dynamic facet. Thus, the question is how complexity increases in evolution, which affects the issue of how prediction on biological systems can be accurate and precise.[26] Within evolutionary biology, there are three general lines to characterize the mechanisms for increasing complexity in biological systems: internalist mechanisms, externalist mechanisms, and undriven mechanisms (i.e., theories invoking no driving force at all).[27]

(I) Internalist mechanisms for the direction of evolutionary change are advocated in five main positions: (a) invisible fluids, which authors such as Jean-Baptiste-Pierre-Antoine de Monet (1809), chevalier of Lamarck,[28] considered present in the environment and kept in constant motion by the sun's energy; (b) the instability of the homogenous, when scientists such as H. Spencer (1890) liked to think that dynamic systems tend to become more concentrated and heterogeneous as they evolved; (c) the repetition and differentiation of parts, a view defended by E. D. Cope (1871), where evolution occurs by the acceleration of ontogenies and terminal addition of parts; (d) the path of least resistance, the conception hold by P. T. Saunders and M. W. Ho (1976, 1981), where component additions are easier to achieve in development than component deletions; and (e) complexity for entropy, where the rising complexity has something to do with the second law of thermodynamics, the position maintained by J. S. Wicken (1979, 1987).

(II) Externalist mechanisms as central for the direction of evolution are, at least, assumed in three versions of evolutionary change: (i) selection for complexity, where the addition of parts permits more division of labour among the parts, and then complex organisms more efficient, according to B. Rensch (1960); (ii) selection for other features, where complexity may increase passively as a consequence of natural selection for other characters, and selection for large size can permit increases, which is the orientation supported by M. J. Katz (1987); and (iii) niche partitioning, where as organismal diversity increases, niches become more complex, which is a conception endorsed by C. H. Waddington (1969).

(III) Undriven mechanisms for the direction of evolutionary change also have three main options. First, *random walk* is when two mechanisms can produce any sort of evolutionary trend, both requiring no driving force at all, such is the view accepted by D. C. Fisher (1986). Second, *diffusion* is when if the initial mechanisms were simple, then later ones could only have been more complex, which is what J. Maynard Smith thought (1970).[29] Third, the *ratchet* is when major evolutionary jumps in complexity occur occasionally in the adaptive radiations accompanying the invasion of new habitats, which is the orientation subscribed by G. L. Stebbins (1969).[30]

These three general lines on mechanisms (internalist, externalist, and undriven), and the set of possibilities pointed out here, need to be tested by the evidence available.[31] Nevertheless, from an evolutionary perspective – an adaptive one – there is a common feature in these three general options of mechanisms for increasing complexity: they are thought of in explanatory terms rather than in predictive terms. Although they can be projected to the future as possible trends, according to our knowledge of the variables at our disposal, the vision is more in tune with a search for an explanation (with the possibility of certain retrodictions) than a clear-cut search of a predictive knowledge open to prescriptive orientations.

On the one hand, these three general lines can use prediction as a possible test, because the evolutionary change looks at the future. Thus, insofar as biological sciences are *basic sciences*, besides explanation they require the use of prediction. On the other hand, biological sciences are also *applied sciences*.[32] Consequently, the knowledge of the future regarding biological systems is needed for problem-solving in specific areas (e.g., the solution of problems such as the possible extinction of species or the elimination of expected consequences of the climate change that can be damaging for nature or for human beings). Furthermore, that biological sciences, in general, and biomedicine, in particular, need predictions and prescriptions has become obvious in recent decades.

Prediction in the Case of Biological Systems

Undeniably, the existence of complexity in biological systems raises problems for scientific predictions in the biological sciences, both as basic sciences and as applied sciences. The variety of the forms of complexity pointed out here – structural and dynamic, epistemological and ontological – creates serious difficulty for the predictive proposals in the biological sciences. At least since Charles Darwin's study of *The Origin of Species by Means of Natural Selection*, the philosophico-methodological analysis of prediction in the biological sciences has been somehow controversial. Two issues are at stake here: the historical consideration and the thematic reflection.

From Explanation to Prediction in the Biological Context

Historically, there is still certain controversy on the predictive character of Darwin's contributions to evolutionary biology. Michael Ruse addresses this point: 'complaints are always made that Darwinism is not truly predictive. You cannot say what will be the future of the elephant's trunk or the giraffe's neck. This is true, but it is to take too-narrow a perspective on prediction'.[33] Thus, if the analysis considers only prediction of novel facts in ontological terms and the remarks concern mainly to *The Origin of Species*, then Ruse's claims are correct. This is the case insofar as Darwin's approach – evolution based on natural selection as the main mechanism for the transmutation of species – was not conceived either with explicit general predictions of new species or with specific predictions on details regarding concrete changes in the species.[34]

Moreover, Ruse is right when he connects prediction with novelty in his text: 'one can and does make predictions – predictions that are successful – for instance, about the ways in which bacteria and viruses will develop defenses against substances that initially are fatal to their well-being'.[35] However, Ruse seems to me wrong when he connects explicitly prediction with the past: 'one can also make predictions (sometimes known as retrodictions) about the ways in which natural selection acting in the past led to expected effects'.[36] The reason seems to be that he confuses 'prediction' and 'retrodiction', which are quite different concepts from a philosophico-methodological perspective.[37]

Prediction is a process and a result – a statement – related to novel facts in three central aspects: ontological, epistemological, and heuristic. Thus, the *prediction* can be connected with the novelty of the reality (a new possible entity or an aspect of an entity), of the knowledge (e.g., an existing reality that we were not aware of), or of the ways of searching scientifically regarding a possible future.[38] Meanwhile, *retro-diction* looks at the past instead of seeking a direct relation to novelty (i.e., ontological, epistemological or heuristic novel facts). When retrodiction states something of the past (either recent or long time ago), the search is, in principle, for elements that might be useful for giving an explanation (what seems to be implicitly recognized through the interest in the expected effects in the past).

If we move from basic science – where the relation is between explanation and prediction – to applied science, where the focus is between prediction and prescription, then we can see again that 'prediction' is conceptually different from 'retrodiction'. The distinction appears in applied science insofar as prescription has a key role: it gives the patterns to solve concrete problems within a specific realm. This step requires a previous prediction, which offers the knowledge about the possible future according to the information available.[39] Consequently, retrodiction cannot be useful for prescription, insofar a statement about the past does not offer us, in principle, any relevant information about the future whose problems we try to solve by means of the prescription.

Certainly, Darwin was interested in retrodiction when he tried to give a mechanism to explain the evolution of the biological systems that he knew (e.g., the relations between the bones of the whales and the reptiles). This is different from the anticipation of the future possible in ontological, epistemological or heuristic terms. Moreover, a large number of biological studies deal with explanatory accounts – either in historical terms (such as the tree of life) or in a contemporary research – of the biological systems available. The biological publications that deal with problems of prediction in an straightforward way is comparatively small. In addition, the philosophical papers on biology are commonly devoted to explanatory issues rather than to predictive topics.

Nevertheless, we need to move from explanation to prediction if we want to enlarge the biological knowledge, in general, and its applied side, in particular. In this regard, it seems to me that we can distinguish with Wesley Salmon between *predictive content* and *predictive import*. It might be that many biological statements have no predictive *content*, but still they may have a predictive *import*. For Salmon, 'statements whose consequences refer to future occurrences may be said to have predictive content; rules, imperatives, and directives are totally lacking in predictive content because they do not entail any statements at all. Nevertheless, an imperative – such as 'No smoking, please' – may have considerable predictive import, for it may effectively achieve the goal or preventing the occurrence of smoking in a particular room in the immediate future'.[40]

Continuing with this distinction, even if some biological statements are lacking in predictive content (e.g., in evolutionary biology), they may have *predictive import* (i.e., corroboration in some cases can provide the basis for deciding which biological theory – with its predictive content – is to be used to make practical predictions). This distinction – predictive content and predictive import – can be helpful to rethink biological statements open to the future. Furthermore, prediction can be in biological sciences insofar as they are basic sciences – they search a systematic enlargement of the knowledge available – and prediction has a role in biological sciences as applied sciences. In this regard, prediction can foretell some traits of the biological systems from the point of view of its structural complexity, and it can also say in advance certain dynamic characteristics of the systems that evolved over time.

One of the persistent topics in this realm is the role of randomness in evolutionary biology and, consequently, how to predict features on the phenotypes and the genotypes in study of biological systems. Following recent research, we can see that 'mapping genotypes onto phenotypes for macromolecules is becoming increasingly feasible. It involves computational predictions of structure phenotypes but also the analysis of existing data on the sequence, structure, and function of ten of thousands of molecules'.[41] Moreover, Andreas Wagner sees here relevant structural and dynamic aspects: 'metabolic networks, regulatory circuits, and macromolecules are important for the formation of most phenotypes: they allow us to classify and compare both genotypes and phenotypes, and they permit us to infer or predict phenotype from genotype to some extent. These three features make them ideal to ask whether phenotypic change is random or not'.[42]

Difficulties in Prediction of Novel Facts in the Biological Realm

Thematically, the historical difficulties of prediction of novel facts in the biological realm and the existence of complexity of the biological systems can be connected with the issue of the obstacles to prediction. The existence of evolutionary novelty seems clear,[43] where the roots in many cases may be in the emergent properties. These are the properties of a biological system that cannot be reduced to the properties that belong to its constituent parts at lower levels.[44] From an overall perspective, the set of obstacles to prediction can be understood based on the possibilities recognized by Rescher for general ontological and epistemological aspects. Thus, he presents them as the *principal impediments* to predictability in any science, while focusing on basic sciences of nature:

(1) Anarchy (i.e., lawlessness or absence of lawful regularities to serve as connecting mechanisms); (2) volatility (i.e., absence of nomic stability and of cognitively manageable laws); (3) uncertainty (i.e., the lack of information about the operative mechanisms); (4) haphazardness (i.e., the lawful linking mechanisms do not permit the secure inference of particular conclusions), which leads to *chance* and *chaos*, *arbitrary choice*, or change and innovation (i.e., outcomes

are not foreseeable because prediscernible patterns are continually broken); (5) fuzziness (i.e., data indetermination whether individually or in a collectively conjugate way); (6) myopia (i.e., data ignorance in the sense of lack of sufficient volume and detail to be able to make a prediction), and (7) inferential incapacity (i.e., the infeasibility of carrying out the needed reasoning).[45]

Noticeably, these impediments to predictability have a relation with complexity, and they pose difficulties to a methodological universalism,[46] which is particularly relevant in biological sciences, where evolutionism has a dominant methodological role.[47] Some of these impediments are mainly structural, whereas others are clearly dynamic. Rescher recognizes that, for many writers, 'complexity is determined by the extent to which chance, randomness, and lack of lawful regularity in general is absent'.[48] But this concept – the inverse of simplicity – is an issue of *degree*: the system can be more or less complex. In the case of biological sciences, the tendency is to focus on some of the previous impediments to predictability, where uncertainty and haphazard have, in principle, a relevant role from the methodological point of view.

Initially, we can point out three main problems for predictability of biological systems: (i) the possibility of a reliable prediction in biological sciences that may be able to deal with complexity in nature; (ii) how to get 'novel facts' in biological sciences (either ontological, epistemological or heuristic novel facts); and (iii) the limits of the predictability in biological sciences, taking into account that, at least sometimes, the whole is more than the sum of the parts (and, therefore, than any biological system may be more than the set of its components).[49]

Unquestionably, the possibility of a reliable prediction in biological sciences is clear insofar as there is an adequate knowledge of the variables at stake regarding the object studied and the problem discussed. In principle, the cases in the short term and in a specific realm (such as bacteria or viruses in an organ) seem more feasible than predictions in the long term and with the widest realm (such as the ecological system as a whole in one hundred years). This can be seen in the controversies on the predictions of the climate change,[50] when they consider the Earth as a whole and a large number of years from now.

How to get 'novel facts' in biological sciences is not easy if they are understood in ontological terms,[51] such as new species in a number of years to be determined. It is less difficult if the novel facts are conceived in epistemological or heuristic terms, such as the discoveries made in recent years with some species in the abyssal waters of the Pacific or new ways to address the research on some diseases of animals or plagues in nature. In principle, when the level of complexity of the phenomenon at stake increases, either because its structure or due to its dynamic, it is more difficult to be able to offer novel facts.

The limits of the predictability in biological sciences can be considered in two main directions: 'boundaries' and 'confines'. On the one hand, there are limits to biological sciences themselves because the topic considered is beyond biology and, therefore, belongs to a different field (such as physics or chemistry, but also other disciplines: sociology, psychology, etc.). On the other, there are limits to the predictability in biological sciences due to the number of variables to be taken into account and the difficulties in getting reliable knowledge about them.[52] In this case, we can distinguish between 'not predictable' and 'unpredict-able': the former is when we are not able to predict in the present moment what is going to happen in a biological system, whereas the latter is when biologists will be never able to predict scientifically such phenomenon or event of a bio-logical system.

Prescription in Dealing with Biological Systems

Besides prediction, prescription has a relevant role in biological sciences as applied sciences. On the one hand, it can state the patterns that can lead the biological system to a specific configuration (i.e., an internal organization); and, on the other, prescription can offer some patterns on how the biological systems should evolve over time, especially when we want to achieve certain goals. Thus, prescription can be connected with the structural complexity and with dynamic complexity of biological systems.

The Need for Prescription

Commonly, the main focus of evolutionary biology has been in the sphere of basic science, because the dominant concern has been explanatory until now, and only occasionally has the predictive component had a relevant task. But bio-logical sciences also have a second domain: applied sciences. Thus, they should solve concrete problems in a specific domain, which requires patterns guiding the solution of the problems.[53] Thus, the need for prescription is clear, because applied science requires to predicting in order to give a guide for prescription, and prescription needs to take into account patterns of action for problem-solv-ing. These patterns are related to values of different kinds.

Prescriptions are patterns on realities that exist now or can be in a possible future as well as patterns on the processes to be developed over time to achieve the solutions sought. Any biological prescription, either at the level of micro-biology or in the realm of the ecological level, requires the evaluation of what is good and bad for the phenomenon at stake (i.e., internal or external values, including ethical values). Before that, prediction gives the relevant information on the possible outcomes according to the ways of action that are thought of. Thereafter there is an application of the biological sciences, which is the use of the agents – individuals or institutions – of scientific knowledge available to the case considered.

Biological prescriptions include, basically, the following features: a) in a clearer form than prediction, prescription takes place on a teleological horizon, because it appears directly related to *ends sought*; b) prescription exceeds the epistemological level which supports the predictive methodology and becomes a concept connected to the *direction of action*, an aspect which is more obvious when a planning of biological activities takes place; c) insofar as prescription leads towards an adaptation of the future of a biological system, it is supported by *the base of predictions* given by the biological sciences as basic sciences; and d) there is an *asymmetry* between 'prediction' and 'prescription', because not all biological prediction is accompanied by the possibility of a viable prescription, due to the fact that it is possible to predict biological phenomena which we cannot actually control (i.e., predictions of biological cycles in the long run) and the biological prescriptions themselves are assessed with respect to their plausibility – as better or worse off biological patterns – according to predictions, and thus predictions can be used to evaluate prescriptions.

Patterns as a Guide for Concrete Problem-Solving

Applied biological sciences can be characterized in terms of aims, processes, and results. (I) The *aims* may be in the short, medium or long run, and they can be located at one of the levels of biological reality: from molecules to the ecological system as a whole. (II) The *processes* depend on the aims sought and the acceptance of the means proposed for these ends (such as avoiding a damage bigger than the cure that they thought of). (III) The *results* should be acceptable both in 'internal' terms (i.e., as an actual solution to a problem, such as a possible cure for genetic disease based on the information given by the Human Genome Project, solving a bacteriological plague, or protecting an endangered species) and 'external' terms (i.e., to be socially, culturally, etc., satisfactory or at least tolerable).

The solutions given to concrete problems by the applied biological sciences require some prescriptions in each biological level. There should be specific patterns thought of as solutions (as concrete as possible) to the problems at stake. These prescriptions can be considered in terms of a 'norm' – or, even, a 'law' – if the proposed solutions are generally accepted as the way of solving the problem (e.g. in cases of the contamination of rivers that harm fish or the pollution of the air that harms plants or birds) after being these solutions repeatedly tested by means of observation or experimentation. Meanwhile the prescriptions can be proposed as a 'recommendation by scientists' – or a 'tentative solution' – to the problem if these prescriptions need to be tested in a new scenario (e.g. after a tsunami or an unexpected outbreak of plague) before it is scientifically endorsed.

Although in the solutions of concrete problems there might be differences between the issues of 'types' and the cases of 'tokens', the acceptance of the pre-scriptions as patterns to guide the solution of problems depends to a large extent on the grounds used as the foundation of their *values*. Thus, the grounds for prescription may be of different kinds, among them are three: ontological, episte-mological, and methodological. These three ways of foundation of prescriptions – and other possible modes – require values to be preserved or to be obtained.

(1) Ontological level is when there is a recognizable good to be preserved in the biological realm (e.g. the existence of some kind of fish in a contaminated river). (2) Epistemological level is when the bases are on the knowledge available to give a reasonable solution to the problem. (3) Methodological level is when the discussion is about the procedure itself to solve the problem rather than the good sought or the preference for a kind of knowledge over other kind (where the decision-making can work on the variety of options of probability theory).

Yet prescriptions are commonly contextual, insofar as any prescription depends, in principle, on the context to establish a set of priorities and the steps for implement them. Generally, prescription involves choosing an option from among several possible ones as the main step. Thus, the aims and processes of the applied science can depend on the criteria used to choose the priorities. Conse-quently, there is an artificial element in the relation with the biological systems when prescription offers a solution to the problems at stake.

Here lies the relation between applied science and application of science, which is particularly relevant in biological sciences in order to deal with biologi-cal systems. In this regard, the kind of knowledge of applied science is different from the application of knowledge,[54] which is the use made by the agents of the applied knowledge within the diversity of settings and circumstances.[55] In addi-tion, we need technology to measure the biological systems (either molecules or ecosystems) as well as technology for transforming such biological systems if a danger for the nature, people or society is involved.

The Role of Technology for Measurement and Transformation

Both epistemological and methodological components of prediction and prescrip-tion in biological sciences require the use of technology. The role of technology can follow two important directions: measurement and transformation. The first has to do with the task of measuring the biological reality available, which should work on structural aspects (mainly, when there is high level of complexity) as well as in dynamic components (which are crucial in any evolutionary study). The sec-ond is the undertaking of technology as instrument in the transformation of the biological reality according to some prescriptions (such as the protection of some natural environments) and, consequently, according to a set of values.

Technology is needed for the measurement of the micro, *meso*, and macro levels in biological systems. This can be considered in three relevant cases. (1) Molecular biology requires the use of microscope and other technological instruments to provide the information, both structural and dynamic, of the reality researched. The knowledge is reliable if the microscope and other instruments are adjusted to the reality studied. (2) Zoological research in animals of a given environment (such as lions in Kenyan savanna) demands the use of technological instruments, such as devices to measure the speed of their movements, the changes in the tone of sounds emitted according to the diverse circumstances (defense of the offspring, attack to obtain food, control of the territory, etc.) (3) Research on the ecological system of the Earth requires specific technological instruments, where the measurement of the biological variables is commonly intertwined with the knowledge of physical and chemical variables.

Observation and experimentation in biological sciences need technological instruments directly prepared for the purpose of measurement of the aspects sought. This involves the technological artifact being made using three facets of knowledge: (a) scientific knowledge (*know that*) from biological sciences and other kinds of scientific knowledge, (b) specific technological knowledge (*know how*) about the instrument as designed artifact, and (c) evaluative knowledge on what is relevant or worthy for this research when there are other objectives (*know whether*, which includes evaluative rationality or rationality regarding ends).[56] Thus, technology can measure biological systems by seeking the relevant variables for prediction and prescription.

In addition to measurement, technology can have the role of transformation of the biological system. Again, the transformation can be in the micro, *meso*, and macro levels of the biological scale (e.g. genetically modified organisms, animal cloning or sustainable development).[57] Here the interplay between science and technology is very strong, as can be seen in biotechnology.[58] On the one hand, scientific knowledge is needed for the technological designs: the prediction should be available to know in advance what might happen; and, on the other, technology is needed for the implementations of scientific prescriptions, either in 'negative' terms (e.g., to eradicate a widespread virus of an obnoxious plague) or in 'positive' terms (e.g. to increase the number of animals of an endanger species).

Undoubtedly, the role of technology for measurement and for transformation is more difficult – and, therefore, more sophisticated – when the complexity of biological system at stake is big. Thus, the attempts to measure our ecosystem and to transform it when needed (e.g., to avoid problems to some species because of the predicted climate change) are more difficult than other uses of technology in a different scale of objectives, processes, and results. The increase in the degree of biological complexity requires a deeper level of artificiality embodied by the use of technology.

Consequently, if we think of our ability to transform living organisms, the lowest level of artificiality is selection and breeding in a small scale and in the short term. 'The next level is brought by Mendelian hybridization technology, operating still through the whole organism in reproduction, but, nevertheless, at the same time focusing on the gene-chromosome (cellular level) as the unit of genetic transmission. Finally, with the arrival of DNA genetics and biology at an even deeper level of understanding, the technology it engenders operates at a correspondingly far deeper level, namely molecular level, thereby generating an even greater level of artificiality in its end products'.[59]

Philosophically, the interaction between biological sciences (from molecular biology to ecology) and the technologies available can be seen form an 'internal' perspective or from an 'external' viewpoint.[60] In the first perspective scientific predictions (at each biological level) should precede scientific prescriptions (either in the 'negative' direction or in the 'positive' direction). Commonly, these are incorporated in technological designs oriented to solving biological problems in nature. From the second viewpoint, society can have a set of values that may be used to establish some priorities in the use of the technological possibilities available. This avoids the risk of a 'technological determinism',[61] as if it were just one option that should be obeyed by the society related to the biological problem.

Following this dual analysis – internal and external – we have a bi-directional relation between biological sciences and the use of technology for measurement and transformation. On the one hand, epistemological and methodological components of prediction and prescription require the use of technology. It is needed for the tasks of measuring the biological reality its complexity, both its structure and dynamics. On the other hand, biotechnology needs prediction and prescription for the creative transformation of the biological systems in the micro, *meso*, and macro levels of nature. This transformation is commonly made according to some social decisions.

To sum up, complexity – structural and dynamic – in biological systems has favoured the preference for explanation over prediction in the biological sciences. In addition, complexity is one of the reasons of the overwhelming interest in biological sciences for the content as basic sciences (the 'theoretical' or 'descriptive' branch) over these disciplines as applied sciences, where prediction serve as a guide for patterns of prescription to problem-solving in concrete cases. The role of technology is dual here. On the one hand, it contributes to measurements that are useful for explanations and predictions of biological systems; and, on the other, technology transforms the biological reality according to the contents of biological sciences as applied sciences (i.e., prediction and prescription) and the contextual criteria of the agents – individual and institutional – that applied these sciences.

9 TELEOLOGY AND MECHANISM IN BIOLOGY

Marco Buzzoni

In spite of a certain oversimplification, one might discern two main opposite trends of thought on the status of teleological claims in biology. Traditional views asserting the intrinsic teleology of living beings proved meagre in scientific results, but it cannot be denied that most biological concepts are defined in functional and teleological terms and Darwin himself used explanations which he claimed revealed the final cause of this or that structure or behaviour.[1] Thus, some authors have also in recent times maintained that reduction of biology to physics is impossible because organisms have a genuine teleological constitution. The teleology of organisms would be intrinsic and not merely projected onto them, in the sense that 'the overall systemic good of an organism is a property of it which has a central place in *explaining* the parts and processes that constitute an organism'.[2]

However, the prevailing trend of thought in the philosophy of biology has consisted in trying to demonstrate that teleology of organisms is only apparent, in the sense that it can be fully explained by empirical forces which are not intrinsically teleological. An interpretation of teleology in naturalistic terms is to be found not only in logical empiricists,[3] but also in the most, if not all, versions of the etiological theory of proper functions.[4] In fact, the most authors would agree with August Weismann's claim that the 'philosophical meaning' of Darwin's theory lies in the fact that it is founded on a principle 'that does not act purposefully, but nonetheless brings about what is suitable for an end'.[5] One of the extreme interpretations of this claim is to be found in Dawkins: 'Natural selection, the blind, unconscious, automatic process which Darwin discovered, and which we now know is the explanation for the existence and apparently purposeful form of all life, has no purpose in mind. It has no mind and no mind's eye'.[6] Dawkins based his definition of biology on this idea: biology would be 'the study of complicated things that give the *appearance* of being designed for a purpose'.[7]

At first sight, the 'new mechanistic philosophy'[8] does not fit neatly into any of the above categories of this oversimplified schema. But it is not difficult to see that it represents a more cautious instance of the second trend of thought. The new mechanistic approach has taken up again, in a new context, the very old and successful attempt to analyse life in terms of mechanisms, dispensing with the explicit reductionist strategy of nineteenth century mechanism and the positivist unity of science.[9] But the question is: is this sort of 'agnosticism' as regards the mechanical world view in its historical prevailing sense radical and consistent enough? In fact, the substantial silence of the main exponents of the mechanist approach on teleology is a clear hint that final causes are thought to be, in last analysis, useless and/or redundant in biological investigations.

After a brief presentation of the mechanical approach in the first section, in the second section of this paper I will show that the concept of mechanism is one-sided and incomplete without a *methodical* (not ontological) reference to final causes. For this purpose, it will be necessary to make a short detour and consider the concept of mechanism or machine in the perspective of the manipulative theory of causality. This will lead me, in the third section, to take a step continuing in the direction of some authors who have tried to reconcile teleology and mechanism, investigating if a third position may be developed, according to which a more satisfactory account of biological practice requires elements of both teleological and mechanical viewpoint, rather than a choice between the two.[10]

As I shall maintain, the 'mechanistic' and the 'teleological' point of view are not reducible to one another; on the contrary, they are intimately connected. On one hand, teleology is not only a necessary assumption in a *general-methodical* sense, but also a *particular-methodical* tool, which appears heuristically useful and powerful in biological practice.[11] On the other hand, there can be no experimental science without the 'principle of the mechanism of nature' (Kant): any conjectured relationship between biological variables could not be regarded as scientific if it could not be tested by the usual experimental method.[12] Both teleological conceptualization and mechanical explanations are therefore necessary for biology, though in very different senses.

Mechanism and Teleology

The fast-growing literature on the mechanist approach has generated a large number of contrasting opinions on many specific points, even though the three approaches prominent in the literature agree on some general theses. For Machamer, Darden and Craver, '[m]echanisms are entities and activities organized such that they are productive of regular changes from start or set-up to finish or termination conditions'.[13] According to one of the definitions given by Glennan, a behaviour mechanism 'is a complex system that produces that behavior by the interaction of a number of parts, where the interactions between parts can be characterized by direct, invariant, change-relating generalizations'.[14] Finally, Bechtel defines a mechanism as 'a structure performing a function in virtue of its component parts, component operations, and their organization'.[15] Apart from some minor differences, recent mechanistic approaches agree in conceiving a mechanism as something analogous to a machine: it is a device consisting of interrelated parts that, starting from an initial situation and ending with a final result, performs some kind of work according to regular, and predictable changes, given like conditions.[16]

In the tradition of classical mechanical philosophy, the attempt to explain life in terms of mechanisms has always played a very important role. However, so far as I know, it was Rom Harré – whose relevance for the mechanist approach is still to be properly appreciated – the first who introduced the concept of mechanism in philosophy of science without the metaphysical ballast of modern mechanism and determinism. Against Hume's approach to causality and the deductive-nomological model of explanation, he maintained that a causal explanation is, after all, the construction of a 'generative mechanism', which is 'responsible' for the connection of a certain cause and a certain effect. A causal explanation, he said,

> consists of certain components which interact according to certain laws, and we find or propose certain conjunctions of phenomena, say p, q and r, within M, which account for the way in which when it is stimulated by a-type events it produces or generates b-type events. Causal relations between these are filled out by describing other mechanisms, M1', M2' ...[17]

As Harré, Mackie and Salmon also developed a theory of causality and explanation, in which a focus on mechanisms is considered as the most promising alternative to the deductive-nomological conception of explanation. As usually recognized, Mackie's and Salmon's analyses of causality directly influenced the new mechanistic systems approach.[18] Since many life scientists – as well as philosophers of biology – had maintained that biology does not possess universal theories comparable to those of physics, the most typical representatives of the new mechanist approach – such as Stuart Glennan, William Bechtel, and the team of Peter Machamer, Lindley Darden, and Carl Craver – refrained from accepting the deductive-nomological model of explanation and were rather inclined to explain a phenomenon 'by identifying and describing the mechanism responsible for it'.[19]

It is certainly true – as Nicholson has rightly noted – that, although the most typical representatives of the new mechanist approach have concentrated their attention upon explanation and causality in biology, they have not confined themselves to these problems and have also tackled further classical topics in philosophy of science, such as reduction, models vs. theories, reasoning in discovery, underdetermination, the nature of laws vs. generalizations, etc.[20] But when all this is admitted, it remains true that they focused on causal explanation, without explicitly dealing with the central question of teleology in biological practice. This applies also to the relevant secondary literature, which mostly neglected the question of the relationships between mechanical explanation and teleology.[21]

The substantial silence kept on teleology[22] or the explicitly expressed fear for the appeal to top-down causes or interlevel causes in biological explanations[23] show that teleology and final causes are thought to be, after all, only useless and/ or redundant information, which should be *de facto* removed.

As we shall see, we may admit that this is true in a sense; but in another sense, fundamental for the understanding of mechanisms and machines, this is utterly false. As for example Polanyi noted, machines (or mechanisms) cannot be understood without taking into account the aim for which they were designed. As Polanyi wrote, machines

> can be recognized as such only by first guessing, at least approximately, what they are for and how they work. Their operational principles can then be specified further by technological investigations. Physics and chemistry can establish the conditions for their successful operation and account for possible failures, but a complete specification of a machine in physical-chemical terms would dissolve altogether our knowledge of the machine.[24]

We identify a machine or a mechanism by its operational principles, that is by the purpose for which it is designed and for which its performance is to be evaluated. Strictly speaking, this would be already enough to show that the attempt to describe and define a mechanism (or a machine) merely as a system that changes regularly from initial to final conditions is doomed to fail.

You might object that 'mechanism' is not to be equated with 'machine'.[25] To this I reply in two ways. First, so far as I know, the main exponents of the mechanist approach have never made a clear-cut distinction between 'mechanism' and 'machine'. On the contrary, they emphasized the connection between mechanisms and machines. For example Thagard[26] defined 'mechanism' as 'a system of parts that operate or interact like those of a machine, transmitting forces, motion, and energy to one another' (which is a typical definition to be found in any English dictionary); Glennan distinguishes two senses in which the term 'mechanism' is commonly used, but a reference to the concept of machine is involved in both cases:

> The first sense refers narrowly to the internal works of machines, as when one speaks of a clock mechanism. The second refers more generally to complex systems analogous to machines, as when one speaks of a human perceptual mechanism or a market mechanism. My analysis can be summarized by a definition which is meant to capture this latter usage: (M) A mechanism underlying a behavior is a complex system which

produces that behavior by the interaction of a number of parts according to direct causal laws.[27]

In Machamer, Darden and Craver 2000, there are not only some historical remarks on this point (cf. section 5), but there is also an important passage, which deserves a quote:

> When a prediction made on the basis of a hypothesized mechanism fails, then one has an anomaly and a number of responses are possible. If the experiment was conducted properly and the anomaly is reproducible, then perhaps something other than the hypothesized mechanism schema is at fault, such as hypotheses about the set-up conditions. If the anomaly cannot be resolved otherwise, then the hypothesized schema may need to be revised ... Mechanism schemata can be instantiated in biological wet-ware ... or represented in the hardware of a machine.[28]

Second and more importantly, there are certainly differences between mechanisms and machines in many contexts (for example, one might say that machines are made of mechanisms), but, as regards the question concerning the relationship between mechanism and teleology, the differences are in degree and not in principle. In principle, as I shall try to prove in the next section, there is an intimate epistemological connection between 'machine' and 'mechanism' on one side, and causality and experiment on the other.

Teleology, Experiment and the Manipulative Conception of Causality

We have seen in the preceding section that, contrary to the notion of mechanism as defended by the mechanist approach, a 'mechanical' investigation of life mechanisms (or machines), while of fundamental importance in the scientific practice of biology, cannot get rid of teleology and final causes. This seems to suggest the thought that final causes are, at least in a fundamental *epistemic* sense, the condition of possibility of mechanical ones. We must though consider also that any conjectured relationship between biological variables could not be regarded as scientific if it could not be tested by the usual experimental method. This tension raises a new problem: whether it is possible to investigate teleology scientifically, taking this word in the sense of Galilean science, without accepting anthropomorphic final causes, which are irreconcilable with modern science.

In order to find a solution, it is necessary to make a short detour and consider the concept of mechanism or machine in the perspective of the manipulative theory of causality.[29] This is no place to enter into the causality problem. I shall assume, therefore, an agency theory of causation, relying in part on Woodward's analysis and in part, to the extent that I do not agree with this author, on my previous account of this issue.[30] However, as opposed to Woodward, I will clarify that the teleological element is a necessary condition of the possibility of the experimental method as we know it from the Galilean revolution.

Woodward has been the first to defend an interventionist treatment of mechanisms. In contrast to geometrical/mechanical views, he has maintained that mechanisms are connected with any interventionist theory interpreted as a 'difference-making account' of causality. According to Woodward, causation and causal explanation, while connected with counterfactual manipulation, are totally free from anthropomorphic elements because they have no special connection with the notion of human agency.

I totally agree with Woodward's interpretation of counterfactuals, which is one of his main contributions to the agency theory,[31] as well as with his and (Price's) insistence that 'our practical interests as agents serve ... to pick out the kind of (independently existing) relationship between X and Y that we are interested in when we worry about whether that relation is causal'.[32] There is, however, a fundamental point of disagreement with Woodward's theory concerning the connection between the notion of human agency on the one side and causation and causal explanation on the other. In the Agency Theory of Causality, the notions of cause and effect depend essentially on our ability to intervene in the worlds as agents. Causes are, as Collingwood said, 'levers' that can be used to manipulate the connected effects, but according to Woodward 'there is nothing logically special about human action or agency: human interventions are regarded as events in the natural world like any other and they qualify or fail to qualify as interventions because of their causal characteristics and not in virtue of being (or failing to be) activities carried out by human beings'.[33]

As opposed to this, I side with some authors, notably G. H. von Wright, H. Price and P. Menzies, for whom the close link between intervention and causality cannot be understood without reference to the free agency of human beings. According to von Wright, for example, the capacity of interfering with the natural course of events is distinctively human and the causal relation depends on the concept of human intentional action.[34] In the same sense, Price and Menzies write: 'an event A is a cause of a distinct event B just in case bringing about the occurrence of A would be an effective means by which a free agent could bring about the occurrence of B'.[35]

My main argument for this thesis is based on a necessary presupposition of scientific experimentation. Following Kant's suggestion, experimentation may be considered as a 'question' put to nature.[36] The answer to the experimental question must be discovered by applying the method of deliberate and systematic variation which requires an active intervention on natural processes. Scientists in principle perform *free* repeatable actions and they *intentionally* modify independent variables in a way that is reproducible in order to determine the consequences of these modifications on one or more dependent variables. More precisely, on one hand experimenting involves an 'external' realization – the construction of an 'experimental machine' –, which extends the original operativity of our organic body, and, thereafter, develops independently of the subject and 'impersonally'. On the other hand, in an experiment, an initial *free* action is necessarily presupposed, guided by the conception of a (cognitive) end. Apart from teleology, apart from the intentional and conscious planning of an experimental set-up and apart from the human actions which freely start or 'set in motion' the experimental machine, it would be impossible to identify causal relations in nature. In a word, there is no scientific knowledge without experimenting and there is no experimenting without our free agency, that is without the free interaction between our body and the surrounding empirical reality according to some cognitive purposes.[37]

To sum up, the manipulative theory of causality presupposes a tacit teleology, namely our free intervening in the surrounding world according to some intentional goal. If we want to explain something scientifically, we have to describe it as if it would or could be the result of our intentional agency. This teleology cannot be eliminated either from the experimentalist or from any other theory of causality *because it is a necessary condition for conceiving natural causal processes objectively* (and not in a naive anthropomorphic sense). Causality is a two-faced coin, one of which – the subjective, conceptual one – is *freely made by agents in order to understand, predict and modify the other – the objective, real one.*

Purposes may be seen as concepts or meanings guiding us in our practical activities. In this strict sense, the concept of an aim or a purpose characterizes only human actions. However, it would be radically false to say that purposes play no part in the explanation of natural phenomena. As far as the mechanist approach is concerned, we conclude, therefore, that the very nature of scientific experiment and the interventionist theory of causality that it presupposes demonstrate that teleology and efficient-mechanical causality are not only compatible, but final causes are in a sense the condition of the *epistemic* possibility of mechanical ones. It is only by reference to the implicit teleology of experimental 'machines' or 'mechanisms' that we are able to give reliable causal explanations.

In biology, as I shall show in the next section, this teleological moment is not only an essential presupposition of any causal imputation, but, as opposed to physics, it is also a general-methical assumption that opens the horizon of biology as well as a powerful heuristic-methodical device to investigate living beings in an intersubjectively testable and reproducible way.

The Counterfactual Use of Teleology in Biology

Ernst Mayr made a very influential attempt to reconcile teleology with mechanism. On one hand, he admitted that it is not only possible but indeed necessary to investigate biological phenomena in biophysical terms. However, he rejected in the most vigorous way the view that the conceptual framework of biology could be entirely reduced to that of the physical and chemical sciences:

> [T]he same event may have entirely different meanings in several different conceptual domains. The courtship of a male animal, for instance, can be described in the language and conceptual framework of the physical sciences (locomotion, energy turnover, metabolic processes, and so on), but it can also be described in the framework of behavioural and reproductive biology. And the latter description and explanation cannot be reduced to theories of the physical sciences. Such biological phenomena as species, competition, mimicry, territory, migration, and hibernation are among the thousands of examples of organismic phenomena for which a purely physical description is at best incomplete if not irrelevant.[38]

To better understand the strength of this argument, we must consider its main epistemological foundation, which consists in the perspectival character of scientific knowledge. As should be obvious after Weber, though it is often forgotten, reality can only be explored from particular points of view which cast light on particular aspects of it.[39] A table can be examined as a physical phenomenon as well as a chemical one. In particular, a mechanical phenomenon results from considering reality from a theoretical and therefore partial point of view, which takes into account only some properties of reality, such as force, mass and certain spatial and temporal relations.

Even if we multiplied the points of view indefinitely, we could never representatively exhaust a real object. A map that reflected a city from all possible points of view would not be a map but a three-dimensional replica of the city. An instrument that was intended to measure an object from all possible points of view would not be able to measure anything: its function consists in allowing us to ignore all aspects of an object except the one or the ones in which we are interested.[40]

This relativization of any scientific discipline, including physics and chemistry, is a serious challenge to any form of reductionism and is sufficient to vindicate the autonomy of biological practice from any other scientific discipline. So far as biology is concerned, one might say that, in a sense that takes up Mayr's argument in defence of the autonomy of biology, the horizon of biology is only given under the presupposition of the notion of organism, seen as a purposive structure which struggles for its survival or organizational maintenance. It is because the courtship of a male animal can be described from this point of view that it falls within the scope of biology.

On the contrary, the answer to an experimental question formulated in physico-chemical terms will only contain physico-chemical concepts; it is for this reason that physics and chemistry never talk about living organisms *as such*. What is an organ from the standpoint of physics and chemistry? These sciences have no concept that could allow us to distinguish between a living being and a lifeless being or between an organ and another organ. In order to be able to make these distinctions, we need the notion of purpose or final cause. To put it simply, each organ is thought of as corresponding to a purpose of an organism: the sight organ is an organ for seeing; the hearing organ is an organ for hearing, and so on.

A qualification must be made here to avert a possible misunderstanding. We must admit that biologists may avoid a particular explicit definition of organism while carrying out their empirical-experimental investigations into the nature and development of actual living organisms; however, without *any* concept of organism, the horizon of a biologist's research would not be given. Whatever may be the empirical-methodical concept of organism used by the biologist at work (and certainly it should be chosen on the basis of its operational capacity to solve specific problems, as Laubichler and Wagner, and Pepper and Herron maintain[41]), it is clear that biology must presuppose the concept of organism in a reflexive-transcendental sense, as a purposive structure that is the condition of the possibility of the methodical point of view of biology as such.[42] But when the field of biology has already been given under the methodical assumption of the notion of organism, only the biologists involved are able to decide which of the empirical concepts of organism is the most promising for their research.

The question now arises: how does biology, given its field of research under the assumption of the notion of organism (and teleology), gain an operational-experimental value, without which biology's scientific character would vanish altogether? It is important not to hypostatize and interpret the reflexive-transcendental concept of finality as an ontological quality. For any attempt to interpret teleology in substantial or ontological terms would place us in an implicitly animistic, pre-Galilean view of nature and would make inexplicable the success biology achieved through the use of a mechanist approach. On the following point I fully agree with the mechanist approach: biologists, after all, have to build a mechanism (or an 'experimental machine') which allows them to intervene technically on reality so as to modify it and then check the effects of that modification according to the well-known 'method of variation' already formulated by Ernst Mach.

The answer to the preceding question is therefore: the assumption of the teleological point of view is fully justified in biology only to the extent that we do not hypostatize this concept as an ontological quality, ascribing to it only a heuristic value and only using it as a counterfactual artifice, capable of bringing to light particular causal relations which have an objective and intersubjectively testable content. On one hand, the purposefulness implicitly presupposed when a living organism is taken into consideration cannot be eliminated or reduced to the mechanical relationship of cause and effect. On the other hand, even though biology is unable either to prove or to disprove the truth of this presupposition, the reflexive, typically human concept of finality may be profitably employed to discover mechanical-experimental causal relations in all living beings. Indeed, project our inner desires and purposes into outer reality is not a mistake in itself. Even though such a projection can lead to the greatest mistakes, it is necessary as a first *methodical* step towards a better knowledge of the living world. Therefore biology may and should substitute where possible the unconscious use of the teleological principle *for a fully conscious and methodical use of this principle*

This thesis may be illustrated by an example. Cells communicate with their external environment through particular channels, that is through proteins that are located in the membrane that encloses the cell and have the capacity of detecting the voltage of the membrane and regulating ion conduction or enzyme activity. In particular, voltage-gated potassium channels alter their conformation in response to changes in the potential of the membrane, thereby allowing or blocking the conduction of ions. Such processes rely on so-called voltage sensors, which are embedded inside the cell membrane and contain an excess of positively charged amino acids which react to an electric field.[43]

If we regard the situation from a predator's standpoint, we shall be able to better understand how the physico-chemical mechanisms of these channels work. In order to do this, it is clear that some counterfactual formulation is required. We might say that if the predator were a human being, it could pursue the aim of crossing the critical threshold of the ion-exchange equilibrium in the prey's cells (according to a line of reasoning which is difficult not to consider a thought experiment, a hint which I cannot pursue here). Which chains of cause and effect should the predator trigger in order to achieve this goal? The predator could, for instance, inoculate the victim with venom that interferes with the mechanisms of voltage-gated ion channels in order to impair the prey's nervous system. Now, this hypothesis, according to which ion channels are blocked by toxins, is clearly suggested by a teleological perspective. However, in order to test this hypothesis, we must use the experimental method. According to the manipulability theory considered above, causal claims are elucidated in terms of counterfactuals about what would happen as a result of consciously planned interventions; however, in order to claim that specific causal chains apply to a particular case, we must have an experimental basis for stating what would happen to the value of a variable if another variable is changed. For instance, it would be advisable (and that is exactly what in fact happened) to test different classes of toxins, whether synthetic or natural, in different circumstances, in order to see whether, how and to what extent they interfere with the external vestibule of the ion conduction pores and work like a plug to block the flow of ions.[44]

As this example shows, *the discovery of certain connections between causes and effects in biology depends upon the heuristic fiction of counterfactually considering nature as if it acted not only purposefully, but also intentionally.* I would even venture to claim that it is an important part of a biologist's work to try to 'empathize' with the living beings that are the subject matter of her or his research, that is to say, to project him or herself into their roles, counterfactually ascribing to them purposes that, strictly speaking, usually we want to ascribe to human beings alone. Indeed, if a thought or a feeling cannot be perceived immediately by the senses alone, an instinct cannot either. One can empathize not only with another human being's thoughts or feelings but also with the instinct of an animal searching for food, with the striving of an organism defending itself against a bacterium in order to stay alive, or with a cell regulating its potassium balance, so that one of the most important conditions for its life is maintained. It is only by putting oneself in the position of a predator, of an organism, of a cell, that it is possible to ask what we would do in similar situations to prey on an animal, defend ourselves from a bacterium or prevent the entry of a toxin which would interfere with the chemical equilibrium that keeps us alive.

It should be clearly understood that this procedure is scientifically harmless since the aim or purpose in question can be bracketed without any loss of objective content. The methodical agnosticism here claimed for biology applies only to the ontological assumption of teleology in nature, not to what can be found to be reproducible and therefore intersubjectively knowable in living beings.

This point is so important that I will illustrate it with another example. According to Konrad Lorenz, just as the horse's hoof is adapted to the ground of the steppe, so also our cognitive apparatus – like any other organ – 'acquired its functional form through ages of encounter of reality with reality'.[45] Even though there is an important element of truth in evolutionary epistemology, there is no doubt that the correspondence between two pieces of a puzzle (or between the horse's hoof and the ground of the steppe) is not of the same type as the symbolic relation of correspondence between facts on one side and the true *or* false sentences which state them on the other.

In spite of this, however, the causal relationship between the horse's hoof and the ground of the steppe can be brought to light in its objective and scientific content *only by putting it in relation to the animal's aim or purpose of fleeing from its predators.* The particular contents of biological research would not come to light as causal-mechanical relationships within the *mare magnum* of the abstract conceivable conditions of a biological phenomenon if they were not seen against the background of an implicit teleology. In this way, the principle of teleology, which – strictly speaking – is probably only valid within the field of human experience, has a proper use outside this field.

That a particular functional form of a horse's hoof is a better (faster and more efficient) way of running can be equally true of an animal that runs away from its predator *in order to save its life by flight* as of a hypothetical hippomorphous robot that simulates this behaviour. In other words, that a particular functional form of a horse's hoof is better for running can be equally true for one who considers animals as complex robots which can be explained entirely through physics and chemistry, as for one who maintains that robots are as intelligent as animals or human beings.

Without the aid of the purposes we project into living beings, there was historically no possibility for biology to have an autonomous subject matter and for physics and chemistry to locate or recognize living beings: biology would be empty, physics and chemistry blind. However, the translation of organic reality into the experimental languages of physics and/or chemistry has a value in itself. It is wholly *independent of the truth as well as of the falsehood* of the 'as-if' teleology of living organisms. Without the translation of organic processes into physico-chemical terms, biology would endorse an anthropomorphic conception of nature. We know that an anthropomorphic description of nature is *always* possible but, exactly for this reason, it must be adopted only as a methodical device to be integrated by means of experimental tests; otherwise, strictly speaking, biology would be devoid of empirical content.

Against Kant, we may say that teleology of living organisms, *though not completely understood*, should not be excluded from the scientific-experimental investigation of biology: a blade of grass may be scientifically investigated if we assume its teleology *counterfactually*, since that makes it possible to exploit it methodically in order to discover the mechanical, and therefore experimentally reproducible, physico-chemical law-like connections at work in it.

We may conclude that, though biology differs methodically from physics (or other experimental disciplines), in the fact that it sees the world from the point of view of 'organisms' (that is entities which struggle for their survival), it does not differ from physics or chemistry regarding its method of discovering causes and effects. There is no special source of knowledge which is open to biology but not to physics or chemistry, and the results obtained by biology is not different as far as their intersubjective value is concerned: they must be obtained through the experimental method, which only ensures intersubjective reproducibility and testability. In the end, biology could not be a science in the historically important sense of this word – that is, in Galileo's sense – should it be unable to express its results in terms of connections of cause (or condition) and effect, and therefore in terms that are experimental in the same sense as that of chemistry and physics.

Even though we may repeat, *mutatis mutandis,* Croce's dictum and say that there is a sense in which we cannot help calling ourselves mechanists, to deny any role to the teleological discourse in biology is no less one-sided than more traditional forms of reductionism and fails to relieve biologists from their discomfort in using a teleological vocabulary. The methodical 'agnosticism' here sketched is able to recognize to teleology a general-methodical function in opening the field of biological research and a more particular-heuristic role in formulating and testing biological hypothesis, where both are only a means for an end, that is for 'mechanistic', experimental, explanations.

10 SCIENTIFIC UNDERSTANDING AND THE EXPLANATORY USE OF FALSE MODELS

Antonio Diéguez[1]

From Droysen and Dilthey until today, it has been common practice in the philosophy of social sciences to consider the notions of 'understanding' (*Verstehen*) and 'explanation' (*Erklären*) as alternative and even opposite scientific methodological approaches. The historicist and hermeneutic schools held that natural sciences look for causal explanation of phenomena, whereas human and social sciences (*Geisteswissenschaften*) should instead look for the understanding of the meaning of human actions and social phenomena in general. Understanding was viewed in this context as a form of grasping and interpreting the meaningful content of some human actions or mental processes, including the meaning that social actors attribute to their actions. The proposed model for this grasping was the comprehensive reading of a text. It is possible to understand the meaning, the aim, or the motivation of a human action in the same sense we can understand the meaning and the aim of a previously unknown written document. This research method is supposedly restricted to social sciences because there is no meaning, aim, or motivation to be interpreted inside the realm of natural phenomena. The behaviour of physical entities does not seem to be meaningful at all and certainly there is no point in trying to make empathetic sense of it.[2]

Nevertheless, this contrast makes use of a very narrow sense of the term 'understanding'. Not only human products or attitudes (actions, intentions, inferences, ideas, works of arts, discourses, etc.) can be understood, but also objective situations, functions, working mechanisms, relationships, etc. And all of these later items are located among the objects of study of physical and biological sciences.[3] We can achieve a comprehensive insight into the meaning of an action, but also into the homeostatic mechanisms that keep the temperature in mammals constant; or into the DNA self-replication process and the grounds of its semi-conservative character. Thus, despite what has usually defended the hermeneutic tradition, it is also possible to achieve an understanding of natural phenomena, and there is no reason to restrict scientific understanding to social sciences. Indeed, the broad acceptance of this fact has contributed to a renewed interest in this topic in current philosophy of science. There have been in the last few years a number of illuminating attempts to clarify the notion of understanding such as is used in the context of natural sciences. It would not be excessive to say that the discussion concerning the role of understanding in natural sciences, and particularly the assessment of its epistemic virtues, is one of the hot topics in the field.

Now then, in the philosophy of (natural) science, understanding has been usually seen as an aim or a consequence of a (good) scientific explanation, and not a methodological alternative to causal explanations. Philip Kitcher expressed clearly this position when he wrote three decades ago: 'a theory of explanation should show how scientific explanation advances our understanding'.[4] In general, the explicit aim of the majority of the proposed theories about scientific explanation has been to satisfy this requirement.

But there has been an additional important change in the focus of the present discussion about scientific explanation. It has been remarked repeatedly that, in many fields, scientists develop explanatory tasks by means of models, instead of laws, theories, or elaborated arguments. In biological sciences this is a well known fact. Such sciences are model-based with various models frequently used with an explanatory function. And in this regard, a surprising variety of accounts has been proposed to cover all the possible modalities this function can offer when it is carried out by a model: structural explanation, simulacrum explanation, mechanistic explanation, equilibrium explanation, causal model explanation, contrastive explanation, explanation by exemplification, explanation by de-idealization, explanation by concretization, explanation by relaxation, and so on. All of them, of course, have been illustrated with suitable examples.[5]

The polysemy of the term 'model' is one of the main obstacles to a comprehensive approach. In biology, for example, 'model' can designate as diverse things as concrete organisms (e.g. Drosophila melanogaster); material objects representing other entities in a simplified form (e.g. molecular models made out of plastic and metal); paradigmatic solutions to an empirical problem (e.g. lac operon model); theoretical and idealized interpretations of the structure and working mechanisms of some biological entities or processes (e.g. the lipid bilayer model of cell membrane, the key-lock model for the enzymatic action, McArthur and Wilson's equilibrium model for island biogeography, Mitchell's model of oxidative phosphorylation); sets of equations describing some aspect of the behaviour of a complex biological system (e.g. Lotka–Volterra model of interspecific competence, Michaelis–Menten kinematic model to determine the velocity of the enzymatic action, or Levins's metapoblational model); simulating computer programmes (e.g. Thomas Ray's model Tierra in Artificial Life); etc. Moreover, not every model is devised to explain a phenomenon. Sometimes, as in the case of a scale model, its function is merely to illustrate or exemplify features of another entity. In addition to explanation and exemplification, models perform many other functions in science (e. g. they can be used as auxiliary elements in experimentation, manipulation or teaching; as a source of surrogative reasoning; as heuristics for getting new hypotheses or for guiding the analysis of alternative possible scenarios; as tools for prediction; as devices for calculating or for making some ideas more precise; as proofs of the possibility of existence, etc.). Demetris Portides[6] has pointed out that most of these meanings of the term 'model' are linked to the ideas of representation and idealization. But even this lowest common denominator can be accomplished diversely by models.[7]

Interestingly enough, some authors[8] have evidenced that the close link between explanation and understanding is particularly visible and revealing in the explanatory use of models. Given the linguistic character of most scientific explanations, it could even be defended that understanding phenomena is a primary and more direct aim of the model building than explaining. From this point of view, understanding is not a by-product of scientific explanation, but in many circumstances it is only because the model lets us explore and understand the details of a particular phenomenon that a scientific explanation of it can be devised. The mere understanding of the phenomenon through the model might be considered in some occasions as such an explanation, as with some models in population genetics. It could be said that these models explain because they provide a genuine understanding of the phenomena. Understanding might be seen then as the primitive target and explanation would be derivative. But whatever the case may be, it is quite clear that the enhancement of understanding is a basic aim to be pursued when a model is used for explanatory purposes. Whether or not it is feasible in some specific occasions to understand a phenomenon without having an explanation of it, or *vice versa*,[9] understanding is in itself a valuable epistemic goal of empirical sciences and models are a powerful device to meet it.

In order to develop the topic of scientific understanding by means of explanatory models, three relevant questions immediately arise and call for an answer. First of all, it is necessary to have a precise notion of what 'understanding' means in this context. Secondly, while scientific models admittedly contain many false assumptions, it has been held that understanding is 'factive', i.e. that it presupposes or implies the truth of the involved beliefs. It is necessary, then, to clarify to what extent and in what circumstances the 'factivity' of understanding should be maintained. Thirdly, understanding has been accused of been devoid of any epistemic value due to its irremediably subjective character. Therefore, if we accept that the understanding of phenomena is a central aim of the use of scientific models, we are obliged to respond to this objection. In what follows I will expose and assess the main answers these questions have received over the last years, and I will defend a personal approach to the second and third question. The next section enumerates some attempts to define the concept of 'understanding'. I will opt for Catherin Elgin's definition because it gathers in a simple form some important aspects of scientific understanding. In the following section I will defend that understanding is not factive as far as the use of models is concerned, since, indeed, false models are good devices for understanding phenomena. To this effect, I will distinguish four types of false models according to the role that falsehoods play in their explanatory function. In the final section, I will suggest some criteria to decide whether a model provides or not a genuine (not merely subjective) understanding of phenomena. For this purpose, it will be convenient to separate what I call 'contrastive models' from the other kind of models that for want of a better word I call 'representative models'.

What Does Understanding (a Natural Phenomenon) Mean?

It is not a simple task to provide a unifying characterization of scientific understanding. As with models, understanding comes in a number of disparate sorts. For example, a distinction has been made between propositional understanding ('understanding that' or 'understanding why' something is the case) and objectual understanding (understanding a phenomenon, an objective situation, a subject matter, a theory, a mechanism, an event, an action, a state of facts, etc.).[10] This last sort of understanding has awakened a special interest in contemporary philosophy of science and will also be my focal point here.

The difficulty of finding a satisfactory characterization of this complex and varied notion is widely acknowledged. Some even think this is a task doomed to failure. It has been claimed in an influential work on this topic that 'it seems to be impossible to give a single universally valid definition of the notion of scientific understanding'.[11] Nevertheless, this difficulty has not impeded many definitions to be proposed in recent years. Michael Friedman[12] holds that 'science increases our understanding of the world by reducing the total number of independent phenomena that we have to accept as ultimate or given'. For Schurz and Lambert,[13] understanding a phenomenon implies knowing how the phenomenon fits into one's background knowledge. For de Regt and Dieks,[14] understanding is to have an intelligible theory of the phenomenon, that is, a theory T about which scientists in a given context can recognize qualitatively characteristic consequences of T without performing exact calculations. Kuorikuoski[15] thinks that understanding is the ability to make correct counterfactual inferences on the basis of received knowledge and to perform effective actions with them. Strevens[16] contends that a person has a scientific understanding of a phenomenon if and only if she grasps a scientific explanation of that phenomenon. Gijsbers[17] considers that we understand a domain D of phenomena if we know which connections exist among such phenomena. Khalifa and Gadomski[18] prefer to say that understanding a phenomenon lies in knowing the phenomenon and knowing an explanation of it achieved through reliable explanatory evaluation. Wilkenfeld[19] defines it as the capacity for carrying out useful representational manipulation. This list of characterizations is only a sample, but it clearly shows that an agreement about the definition of 'understanding' is far from being reached. Some think that it is a kind of knowledge; others think it's a skill; others that it's a mere subjective experience; others a capacity for something (making new inferences, mental manipulation, constructing models, effective action); etc.

I do not aspire to put my own definition on the map. Instead of increasing the collection of plausible definitions, I think it is better to select among the many already proposed one good enough for our aims. I judge Catherine Z. Elgin's definition[20] elegant, comprehensive, and clarifying, and therefore I will assume it here. This is her proposal:

> [U]nderstanding is a grasp of a comprehensive general body of information that is grounded in fact, is duly responsive to evidence, and enables non-trivial inference, argument, and perhaps action regarding that subject the information pertains to.

Besides its simplicity and accuracy, one of the main qualities of this definition lies in its usefulness to see how scientific models are used in explanatory functions and how this use allows us to understand some phenomena. Actually, it includes both the theoretical and the practical aspect of scientific understanding. On one hand, understanding is described as a mental grasp of information that enables us to infer new interesting consequences about phenomena. On the other hand, this mental achievement makes possible to act on the phenomena – including mental manipulation. According to this characterization, understanding is manifested in a double capacity: the capacity for further reasoning on the phenomena under scrutiny and the capacity for successful manipulation of the same. And, supposedly, both aspects are closely entangled in the practical strategy of model-based sciences.[21]

Understanding (via) False Models

As for the second question – whether or not understanding presupposes the truth of the beliefs held about the understood phenomena –, Jonathan Kvanvig,[22] among others, has defended an affirmative answer. For him, to say that a person understand *p* requires that *p* be true. As I said above, he distinguishes between propositional understanding and objectual understanding. The first lies in understanding that something is the case, or in other words, in understanding a proposition describing something. The second lies in understanding an object (a phenomenon, a situation, a language, etc.). According to Kvanvig, in order to understand a proposition, the proposition must be true; and similarly in order to understand an object, the beliefs about this object must be (mostly) true as well.

However, this claim does not fit well with the frequent use of false models in science. In spite of its current use in philosophical literature, many persons still have misgivings about the expression 'false models'. Models do not seem to be true or false in a strict sense. In the first place, usually they are not linguistic entities, and in their formal and rigorous usage 'true' or 'false' are metalinguistic predicates applicable only to linguistic entities. Secondly, models themselves do not state anything about reality unless they are supplemented by what Ronald Giere[23] named 'theoretical hypotheses'. These hypotheses declare that the real system is similar to the model in some respects and to some degree. They would be true or false, but not the model as such. And thirdly, in general, models present a lot of idealizations (and abstractions) that neglect or distort important aspects of the real world, and make of any model system something unavoidably fictional. Thus, the hypothetical system described by a scientific model usually is unreal and idealized. There are no ideal gases, perfect pendulums, completely isolate populations of predators and preys, or infinite populations of random-mating individuals. They are only useful imaginary hypothetical entities; i.e. fictional constructs created for the sole purposes of research. Of course we can generate some 'fictional truths' with them, and in this sense some propositions are true or false inside the model (e.g. in the Copernican model of the universe, the statement 'planetary orbits are circular' is true). But these fictional truths are not literal truths about the real world, because they do not refer to any actual trait, but to a fictional one.[24] In as much as the model as a whole is just an idealized fictional representation of this sort, it is neither true nor false with regard to a real target system.

Nevertheless, models can be judged true or false in a broader and indirect sense. Similarly to the maps, they can be also interpreted as accurate or inaccurate representations of real-world features. A map is a partial, perspectival, simplified, conventional, historically contingent and indefinitely perfectible representation of a territory and of some its geographical items (railways, restaurants, monuments, etc.). A model is also a representation of something else; it stands for a real target. Once we take into account the current reading conventions and the different aims and interests that the map-maker could have pursued, a good map must show some interesting and contextually useful structural similarities with the territory represented, and must preserve the significant relations between its parts or elements. These preserved relationships make possible that the surrogative reasoning carried out through the map (such as 'If I'm in Toledo, I have to take this road to arrive to Madrid') lead to correct consequences concerning the actual traits of the territory. If this condition is not fulfilled in a satisfactory degree, it is not preposterous to say that the map is misleading, inaccurate, incorrect, or – at least in extreme cases – simply false. A map of Spain in which Madrid is located between Málaga and Toledo, or the Guadalquivir River flows into the Bay of Biscay, would be not only an inaccurate map, it would be a false map *tout court*. Otherwise, if the condition of preserving the significant structural relationship is met, the map could be estimated as roughly accurate or approximately true.[25] In the same way, bearing in mind the purpose for which a model has been put forward by scientists, it is to be expected that such a model provides faithful descriptions of some interesting and relevant properties of the target system in that context. Hence, a very unrealistic, incomplete or counterfactual model, a model that does not share enough relevant properties with its target system or that fails to refer to its real properties could be considered false.[26] In fact, due to the mentioned idealizations constitutive of many models, some authors think that these models cannot but be false. Highly idealized scientific models are false in the sense that they do not – and cannot – offer an accurate representation of the real target system.

This view is in need of further qualification though, for it is well known that at every turn scientists derive from these high idealized models and from other kind of false models a lot of useful and substantial true consequences about the characteristics and behaviour of their targets. The interesting point is, then, that false models can be used under certain conditions to get a revealing insight into the working mechanisms or into the causes of actual phenomena. As it was said before, false models are frequently used as tools for providing scientific understanding. When this happens, they constitute what Elgin[27] has called 'felicitous falsehoods'. An adequate understanding of these false models can yield an adequate understanding of the target systems. Some authors, like Hindriks,[28] Morrison[29] and Kennedy,[30] are more radical and they have argued – convincingly, in my opinion – that it is precisely in virtue of the model's inaccuracies or falsities – and not despite them – that it explains or provides understanding.

I think a helpful step in order to see how these falsehoods can be felicitous ones and so facilitate a scientific understanding of phenomena is to distinguish several types of false explanatory models. Because, actually, models can be false in several different ways and this is not irrelevant to their use. I outline here a classification which does not pretend to be exhaustive, but that could be used as a guideline.[31] I will distinguish between (1) adjustable models (2) template models (3) non-referential models, and (4) contrastive models. Let us dwell a bit upon the differences:

(1) *Adjustable models*: They are false models susceptible of improvement through a progressive process of de-idealization or concretization, leading to describe the phenomena in specific situations with greater accuracy and realism. The ideal gas law ($PV = nRT$) is a good example. It comprises an idealization assuming that molecules are perfect elastic particles with negligible volume and

that there is no attraction or other interaction between them. In the normal range of temperature and pressure this law remains valid, but not in scenarios with very high temperature and pressure. However, the van der Waals's equation introduces two corrections; one to reflect the volume of molecules and another to take into account the attraction between them. This law is then a refinement of the previous one. It contemplates more particular features of the real gases and is valid also for high pressures:

$$(P + n^2a/V^2)(V - nb) = nRT$$

Volterra model for a population of predators and preys could be cited as another exemplary case. It states that the dynamic of interactions between the two species in this type of population can be suitably described by two simple equations. If D is the density of predator population and P is the density of prey population, then the growth rate of prey would be

$$dP/dt = rP - aPD,$$

and the growth rate of predator would be

$$dD/dt = haPD - mD,$$

where r is the instant growth rate *per capita* in the prey population, a is a measure of the rate of capture for each predator, h is a measure of the efficiency in the transformation of the energy obtained from the captured preys in the production of new predators, and m is the specific mortality or migration rate of predators. Anyway, this elegant model makes a number of idealized assumptions. It supposes that the only restriction to the prey growth is the existence of predators, so that in absence of predators the prey population would grow indefinitely. It is easy to correct this false assumption by introducing a new factor in the equations. Let K be the carrying capacity of an environment, that is, the maximum population size of preys that the environment can sustain, then the equation for the prey population growth rate would be

$$dP/dt = rP(1 - P/K) - aPD$$

That new factor replaces an exponential growth for a more realistic logistic growth. And this replacement leads to important changes in the dynamical behavior of the system; namely, the constant oscillations of the populations' size produced in the first system are gradually cushioned until reaching a steady state, or simply they do not appear any more. And a similar correction is feasible in order to take into account the mere fact that any predator has a limit in its capture rate.[32]

(2) *Template models*: They are models describing a non-existent ideal situation from which actual systems deviate to some degree due to the influence of several causal factors to be empirically determined in each case. Hardy-Weinberg's model fits in this sort of models. It is in fact a case of 'neutral model', in

Wimsatt's sense.[33] Neutral models in evolutionary biology are models without natural selection. They try to describe a situation in which natural selection plays no significant role. But Wimsatt extends the idea to any 'baseline model that makes assumptions that are often assumed to be false for the explicit purpose of evaluating the efficacy of variables that are not included in the model'.[34] And he adds: 'the use of these models as «templates» can focus attention specifically on where the models deviate from reality, leading to estimations of the magnitudes of the unincluded variables, or to the hypothesis of more detailed mechanism of how and under what conditions these variables act and are important'.[35] The Hardy–Weinberg model establishes that the gene frequencies of a panmictic and infinite population not subject to natural selection, mutation or migration, remain constant across the generations. Let us suppose that there are two possible alleles for a certain locus in the genetic pool of this population; the allele *A*, with an initial frequency *p*, and the allele *a*, with a frequency *q*. If so, the frequencies in the equilibrium state of the genotype *AA*, *Aa*, and *aa* will be p^2, $2pq$, and q^2 respectively. Virtually, no real population meets these criteria, but precisely what the model aims to bring out is the fact that any deviation from the equilibrium-frequencies founded in the real measured gene frequencies must be explained by the intervention of one or several of the excluded factors (natural selection, mutation, migration, genetic drift due to the small size of the real population, not-random mate, etc.). In that case, it would be pointless to de-idealize the model, for its function is not to understand the working of a real system in simplified situation, but to fix the point beyond which some evolutionary force must have been acting. As Sober[36] explained some years ago, this model describes a 'zero-force state'. Another illustrative example of this kind of models would be R. A. Fisher's sex ratio model. Fisher's model explains why under a variety of conditions the stable (adaptive) proportion of males and females in biological populations is 1:1. If the actual sex ratio in a population departs from this proportion, there must be some cause of this disparity to be found. The populations of certain species of invertebrate parasites, for instance, usually contain a higher number of females than of males. These populations live in isolated habitats, where the mixture with other populations is very unusual, and the individuals have to reproduce very quickly. Then, the mate competition is merely local, e.g. only against the rival individuals staying in a host. Under these circumstances, the action of natural selection favours a female-biased sex ratio. The same thing happens with a polyginous species, the Scottish red deer *Cervu smegalocerus*, in which the more dominant females are better fed and their offspring grow up to become stronger than the average. These dominant females tend to have more sons than daughters.[37]

(3) *Non-denotative models*: They are irremediable false models, since the model postulates entities, properties or mechanisms that fail completely to refer.

There is nothing in the real target system able to be designated as the element denoted by the main elements of the model. For this reason, unlike the two previous kinds of models, they cannot be considered even approximately true. Sometimes, but not always, such models are used only to save the phenomena, like the Ptolemy's epicycles. Other times, like in the case of caloric or the electromagnetic ether, these models are the ones that simply fail to denote the real features of phenomena. This kind of models, inasmuch as they are not identical to other false models, has been comparatively unattended. Three notable and interesting exceptions are Morrison,[38] Elgin,[39] and Toon.[40] Morrison analyses in detail Maxwell's ether model as an illustrative example of how a fictional mechanism can provide information and yield useful predictions. She contends that there are a number of ways these models might accomplish this task, hence a careful analysis is required in each case. Elgin explains that these models are not fictive, like the ideal gas model, but defective. Unlike fictive models – which do not try to denote any real object – these non-referential models purport to denote an actual target system, but in fact this system does not exist in the real world. In the case of fictive models there are real target systems that share with the model some of the properties exemplified by it. This does not happen with the non-denotative false models. They cannot share any property with the target system because this target does not exist at all. For his part, Toon states that not always these models try to represent an actual object; sometimes the intended target system is explicitly a non-existent object, e.g. a model of a bridge not yet built, or a ball-and-stick model of unreal atomic configurations.

(4) *Contrastive models*: With some exceptions in this last case, these three previous types of false models are built to fit somehow with their target, or, in other words, they are models built to represent (some aspects of) a real-world target system. But there is another type of false models not encompassed by the mentioned types. They are models explicitly formulated to represent imaginary target systems. There would be no actual system that could be considered to be represented (even approximately) by them. But although the represented system is a completely fictional one, it is proposed however in order to understand some real phenomena. Sometimes these models can even be in conflict with the accepted scientific laws. And, unlike other models based on ideal non-existent model systems (ideal gas, perfect pendulum, infinite populations), in this case, it is the dissimilitude or the contrast between the model system and the real-world phenomena that has the burden of the explanatory function and casts light on the workings of these phenomena. As Weisberg states, 'insofar as we can understand why do not exist [the phenomena described by the model], we will have gained a better understanding of phenomena that do exist.'[41] For that reason, these models can be designed as '*contrastive models*'.[42] A good example would be – in my opinion – Laurence D. Hurst's[43] model to explain why usually there are

only two sexes in nature. Hurst offered a mathematical model for a population of isogamous protists in which there are three mating types. The model shows, among other things, that unless the costs of finding a mate are high, this population and other populations of organisms with gametic fusion should evolve towards two mating-types or sexes, and therefore, that organisms with gametic fusion and more than two sexes should be rare in nature. More recently, but along the same lines, Tamás Czárán and Rolf Hoekstra[44] constructed a model showing that, under usual conditions, 'a population consisting in two mating types can displace a pan-sexual population which is otherwise similar to the mating types in all other respects.'[45] A species or population is pan-sexual if every sex cell of the organisms which make it up can potentially fuse with any other sex cell. In a similar way, a model developed by Michael Bonsall[46] indicates that, unless that numerous and complex physiological innovations were introduced, diploid zygotes have higher fitness optima than triploid zygotes (i.e., zygotes produced by the fusion of three types of gametes). In these three examples, the propounded models deal with three or more mating-types. With the exception of fungi and a few rare organisms,[47] this is an unreal situation, so that the models do not aim to match any actual system. However, they show that these unreal modeled systems are instable or have lower fitness than a two mating-type system. In this sense, they provide some understanding concerning the widespread existence of this kind of reproductive modality. From the point of view of their explanatory use, these models diverge from others in an important issue; what can be named as their *representational intentions* are not the same. An ideal gas is an idealization of real entities: ordinary gases. As a model, the ideal gas tries to map onto any real gas in a real context. On the contrary, three-sex species are not idealization about anything real, but fictions not accessible by means of simplification or abstraction carried out on real systems.

We are now ready to bring into focus the question about the factivity of scientific understanding. First of all, we must distinguish between understanding a phenomenon by means of a model and understanding the model itself. When we consider the case of understanding *a* model itself, understanding seems clearly not factive. We can obviously understand false models, e.g. Ptolemy's model of planetary motions can be understood by any dedicated student of the history of astronomy. In a similar way, we can understand false propositions, like 'Spain is an island', or propositions that are not literally true, but only 'true in fiction', like 'the books of chivalry drove Don Quixote crazy' or 'all the vortex in electromagnetic ether spin in the same direction'. Furthermore, we can understand what an ideal pendulum is, or what an ideal gas is, although there is no ideal pendulum and no ideal gas in the real world. This is a trivial fact usually admitted by those who defend the factivity (or quasi-factivity) of understanding. What they rather emphasize is that things are not the same when we turn our attention to the understanding of an objective situation or a phenomenon. In that case, it seems that the phenomenon to be understood cannot be a spurious phenomenon. Following Kvanvig's suggestion, it would not be feasible to understand that p is the case if p is not the case indeed. It would be senseless to say that 'John understands why sugar never dissolves in water' or that 'Mary understands how the rain dance produces rainfall'. It also seems that you cannot understand the cause of something if your beliefs about this cause are outright false. You will fail to understand why your car broke down if you erroneously believe that the cause is a defect in the carburetor whereas the real cause is a problem with the spark plug cables. And, what is more interesting here, some of these authors hold that the understanding of a false model does not involve the understanding of the modeled phenomena. 'One might understand – writes Kvanvig[48] – the model or theory itself, as when one understands phlogiston theory. One does not thereby understand combustion, however'.

Now then, it is a very strong position to demand factivity to any form of scientific understanding. *Pace* Kvanvig's statement, sometimes we can understand the behavior of a real target system – like a gas – or imaginary systems – like a three-sex species of animals – by means of models containing a large number of false suppositions. And in this use of models, falsities are neither peripheral nor dispensable. Let us see in a more detailed way how the different false models we have distinguished realize this function.

Adjustable models are idealized and abstract representations of a target system. However, these idealizations and abstractions, as explained, are not impediments to the correct understanding of the behavior of phenomena, rather they are tools to achieve it. Adjustable models allow us, for example, to foresee how the target system would change if some initial conditions were different. They give some relevant answers to the frequent demands of counterfactual information about the behavior of the target system in a variety of circumstances. On the other hand, as Elgin[49] states, these models exemplify some of the most significant properties of the target system and make easier for us the analysis of the relations and interactions of these properties. As she writes,[50]

> No real gas has the properties of the ideal gas. The model is illuminating though, because we understand the properties of real gases in terms of their deviation from the ideal. In such cases, understanding involves a pattern of schema and correction. We represent the phenomena with a schematic model, and introduce corrections as needed to closer accord with the facts. Different corrections are needed to accord with the behavior of different gases. The fictional ideal then serves as a sort of least common denominator that facilitates reasoning about and comparison of actual gases. We 'solve for' the simple case first, then introduce complications as needed.

Thus, adjustable models lead to a better understanding of real phenomena insomuch as the processes of de-idealization and concretization yield to better predictions and analyses. These processes, by contrast, are not so relevant in the case of template models, even though they provide understanding in an analogous way. They are used as a resource to detect the reasons of the departure of real systems from an ideal situation. Insofar as these reasons are figured out, we reach a better understanding of the target-system's working circumstances, as well as of the factors that modify and shape it. With regard to contrastive models, they work in another way, but not very dissimilar. Unlike the adjustable models – and in some sense, also the template models –, where it is the similitude with the real target system what is interesting, in the case of contrastive models, it is the dissimilitude what has the burden of the explanatory function and casts light on the workings of the target system. They answer to the contrastive question 'why p rather than q?', and therefore they let us understand – for example – why there are only two sexes rather than more.[51]

Non-denotative models display more difficulties. Whether or not these non-denotative models are able to yield some sort of understanding is a complex and debatable matter. On the one hand, it could be argued that they do not constitute a basis for a genuine understanding at all. They would be simply cases of misunderstanding. It is true that the Ptolemaic astronomy was useful for navigation and to predict eclipses, but in fact epicycles are not even remotely connected with the actual mechanisms that cause planetary motions. Therefore, they cannot give us a genuine understanding of these motions. And the same would be applicable to the phlogiston chemistry and to Maxwell's ether model. Wimsatt[52] seems to adopt this position when he writes: 'Will any false model provide a road to the truth? Here the answer is just as obviously an emphatic "no!". Some models are so wrong, or their incorrectness so difficult to analyze, that we are better off looking elsewhere'. But, on the other hand, it could also be argued that this kind of models can provide a defective but valuable understanding of the phenomena. At any rate, they are better than simple ignorance. From an instrumentalist or a constructivist point of view, these models have made possible a certain degree of control over the phenomena or have contributed to set up some interesting possible world, so, although they cannot be seen as approximately true representations of reality, they were useful devices in our practical and cognitive handling of the world. Apparently, knowledge and understanding do not follow the same epistemic rules. A false knowledge is no knowledge at all, because knowledge must be true by definition. Understanding, in contrast, does not seem to be a so dichotomous issue. A false understanding (a misunderstanding) can be somewhat a kind of incipient or imperfect understanding, and can be judged in a later period as the first incorrect step in a way leading to the current genuine understanding of phenomena.

Then, a balanced judgment on the role of non-denotative models would demand an extensive historical analysis. Of course, we cannot do it here, but we can remind some well-known few things that will help to contextualize the issue. Ptolemaic planetary models meant a real progress in the understanding of the structure of the universe respect to Eudoxus and Calippus models and to the Aristotelian spheres model. Assuming that one of the main functional characteristics of a scientific model is carrying out surrogative reasoning about its target,[53] we must acknowledge that Ptolemy's models fulfill reasonably well this function when it was put forward. It is probably true – although some historians dissent – that it was proposed as a mere mathematical model to calculate the position of planets, not as a physical model trying to represent the working mechanisms of the cosmos. At least, it was commonly interpreted in this way during the Middle Ages. But Ptolemy's models were able to generate some consequences concerning the changes in the brightness of the planets, their apparent retrograde motions, the variation in their velocity on the background of the ecliptic, the absence of stellar parallax, and so on.

However, from the contemporary perspective, we see this set of models as flatly false representations of reality – not as approximate truths–,[54] and we tend to consider them unable to provide any genuine understanding of planetary motions. They cannot give an answer to most of the questions that might be raised from the present knowledge perspective about these motions. Its representation of them differs completely from ours. We could not accept this representation without refusing almost all of our scientific knowledge about the Solar System and about physics. Thus, few persons would probably be prepared to accept that Ptolemaic models supply some form of understanding of the functioning of the universe.

For its part, likewise epicycles, or phlogiston, or caloric, Maxwell's ether is a fictional entity (although some scientists, like the British physicist Oliver Lodge, believed at some point in its real existence, as was the case with epicycles, phlogiston, or caloric as well). But we are probably less reticent to admit it played a fundamental role in the understanding of physical phenomena. The mechanical model of the electromagnetic ether had an extremely valuable function in the articulation of Maxwell's electromagnetic theory. So, from our current perspective, it is reasonable to think that its contribution to the development and advance of this theory justified its use, independently of its failure to denote a real-world system. It facilitated the calculations, guided ulterior researches, had a heuristic value, and was a very useful device for surrogative reasoning and in deriving some mathematical relationships. It led Maxwell, for instance, to the conclusion that the light must be itself a kind of electromagnetic wave, since transversal waves were transmitted in ether at the speed of light.[55] But Maxwell attributed to it basically an illustrative and auxiliary function, without intending to reflect anything real, and finally his theory – the field equations – dispenses with the model.

What makes so different the way we contemplate Ptolemaic epicycles models and Maxwell's ether model in reference to their role in our understanding of natural phenomena? I think that a good indication to elucidate this question can be found in this reflection by Margaret Morrison:[56]

> Given that many models cannot be evaluated on their ability to provide realistic rep-
> resentations, we need to focus less on the distinction between 'heuristic' and 'realistic'
> models, and instead, emphasize the way in which models function in the develop-
> ment of laws and theories.

A clean-cut difference between Ptolemy's epicycle model and Maxwell's ether model is precisely that the later was a useful device in the development of laws and theories nowadays accepted, but not the former. Accordingly, it could be stated that we understand real phenomena by means of non-denotative models only if these models were useful in the development of laws or hypotheses that could be justified by currently accepted scientific theories.

The (Desirable) Objective Character of Scientific Understanding

The question concerning the objectivity of understanding was controversial from the very beginning of the discussion. Hempel[57] famously held that under-standing is a merely psychological or pragmatic matter; a notion to be attributed only to subjective states of individuals, and consequently a relative and non-gen-eralizable concept. Unlike explanation, an account of understanding necessarily involves a subject. The feeling of understanding, however strong it may be, does not imply a genuine understanding. This thesis has been held by J. D. Trout as well.[58] Trout argues that a 'sense or feeling of understanding' is by no means a reliable indicator of the truth of a scientific explanation and it is neither neces-sary nor sufficient for good explanation. The more significant evidence adduced by Trout in support of his claim are the experiments carried out by cognitive psychologists showing the common human biases due to overconfidence and hindsight mistakes.

An early reply to Hempel's position was formulated by Michael Friedman:[59]

> [A]lthough the notion of understanding, like knowledge and belief but unlike truth,
> just is a psychological notion, I don't see why it can't be a perfectly objective one. I
> don't see why there can't be an objective or rational sense of 'scientific understanding',
> a sense on which what is scientifically comprehensible is constant for a relatively large
> class of people.

More recently, Elgin[60] makes a similar point:

Even though human subjects understand, it is not obvious that their accomplishment should be characterized as subjective. To see this we might note that understanding is closely related to knowledge. Although knowledge involves belief, no one is inclined to say that knowledge is merely psychological, not epistemological. No one holds that whether *s* knows that *p* is subjective. Why should understanding be different? Knowledge is related to justification, which typically relies on tacit background beliefs. But although people may think they know because they consider their justification and background beliefs adequate, they can be wrong, even if they satisfy the standards of their own epistemic community. If knowledge is not keyed to the standards of a particular, historically situated epistemic community, why should understanding be? Why shouldn't we say that our predecessors thought they understood the motions of the planets, just as they thought they knew that the earth was motionless, but in both cases they were wrong?

I have dealt with this problem elsewhere,[61] and I will sum up my position here.

For a start, I think it is important to draw a line between contrastive models and representational models, since they constitute two different strategies to reach scientific understanding. Contrastive models, as we have seen, are false models that allow us to understand a real system by showing why some situations related to this system are impossible or very improbable in normal circumstances. Representational models are models explicitly designed to represent after all a real target system. The other three types of false models we have presented – adjustable models, template models, and non-denotative models – belong to this last class.

This distinction makes it easier to find criteria for genuine understanding. In fact, this distinction is necessary because, in view of their diversity of aims, the criteria cannot be the same for both types of models. These criteria could be interpreted as tentative indications to decide when a scientific model is able to provide genuine understanding, and not a merely subjective feeling of understanding, to an informed individual. For the case of contrastive models, I would now suggest a very simple quality criterion of genuine understanding:

> A contrastive model gives us a genuine understanding of the behavior of the real system if the contrast between the consequences derivable from the model and the real target system can reveal how an interesting characteristic of the behavior of the real system might depend on the presence or absence of certain circumstances which are respectively absent or present in the model, or if the model shows how its unrealistic assumptions are hypothetically unstable and for this reason the opposite conditions prevailing within the real system tend to arise.

Thus, if the entities of the model, or its properties, are unrealistic, but the casual mechanisms postulated by the model are analogous to those of a real system, it would be possible to learn about the limits and potentialities of these causal mechanisms and, therefore, about their operations in the real system. The function of this first kind of model is certainly closer to instruments for exploring the world than to faithful representations of reality.

As for representational models, things are more complex because they are much more used in science and they offer a greater diversity. However, it is possible to pick out some criteria which might be used as indicators of a spurious sense of understanding. Since representational models make some realistic assumptions about the target system, I think these criteria must be focused on the methodological and epistemic resources that could strengthen the reliability of these assumptions. Using Weisberg's[62] terminology, they can be interpreted as minimal 'representational fidelity criteria'.

> A representational model provides a genuine understanding of the target system if:
>
> (1) the analogies between the model and its target are not weak or scientifically unfounded;
>
> (2) it does not formulate oversimplifying abstractions which exclude relevant functional factors, i.e., factors which are necessarily constitutive of the behavior of the target system;
>
> (3) it does not make extremely unrealistic and useless idealizations, that is, idealizations which are so far removed from the real conditions of the modeled system that they do not help to see how the behavior of this system varies under the action of usual causal factors or under certain manipulations;
>
> (4) it does not postulate a pseudoscientific ontology; i.e., it does not postulate entities or processes incompatible with the current state of science;
>
> (5) the postulated mechanisms offer analogies with the mechanisms that are working in the real system;
>
> (6) its predictions about collateral phenomena do not fail systematically.

I think these criteria highlight typical deficiencies which go against the possibility of a faithful representation and increase the arbitrariness of the model. The more these criteria are unfulfilled, the less accurately the model represents the target system. It is reasonable to think that these deficiencies make it more probable that the surrogate inferences carried out with the model are too misleading or uninformative. Detecting some of these deficiencies is, then, a good reason to conclude that an initial sense of understanding caused by the model does not correspond to a genuine understanding of the target system. But if these deficiencies are not detected and the criteria are met, we can be confident that the sense of understanding in such a case is not merely a subjective feeling.

Conclusions

False models are excellent devices to get a scientific understanding of natural phenomena. Taking into account the different ways they can pursue this goal, it can be distinguished between adjustable models, template models, non-denotative models and contrastive models. All of them involve falsehoods which are necessary to the explanation of the behavior of the target system and to the understanding of the nature of real-world phenomena. Therefore, understanding, unlike knowledge, is not factive. It does not presuppose that the majority of the beliefs involved in the state of understanding must be true. Finally, understanding is not irremediably subjective. Some reasonable contextual criteria can be chosen in order to tentatively assess when a feeling of understanding corresponds to a genuine understanding.

11 BIO-TECHNO-LOGOS AND SCIENTIFIC PRACTICE

M. Bertolaso, N. Di Stefano, G. Ghilardi and A. Marcos

According to the opinion of many philosophers of science, science should be made of empirical observations, such as those we find on current scientific journals, laboratory reports and reviews. What makes these works scientific is a peculiar method, completely grounded on observation and logical inference. This is the classical dominant idea of science that seems not to resist any longer. Considering science as a human action will open us different dimensions that characterize scientific work, and that are frequently neglected in the logical approach to science, such as the linguistic aspect of scientific theorization, the ethical and the esthetical dimension, the political sense, and all the sort of things that goes under the 'human factor'. This is apparently a little shift of perspective though it opens up a whole series of dimensions that cannot be kept away from science anymore.[1]

Once it has been made clear the connection between science and the domain of the action, the individual sphere and the social context, we need to move further in order to correctly characterize science and specify that he who does science is always a human *person*, not just a sort of individual monad dropped out of context. This stress on the personal status of those who deal with science leads us to consider that when we admire a scientific work, we are not just admiring the product of few capacities or functions of those who delivered such scientific achievements, but rather appreciating the work of persons who express much more than logical reasoning and observational spirit in their outcomes. Capabilities are those logical reasoning and observation, but obviously, scientific work is much more than that. Science is the whole person, with all its abilities, attitudes and circumstances. This includes emotions, feelings, motivations, interests, attention span, intuition, imagination and memory, aesthetic and moral sense, social context (dialogue) and historical (traditions).

It could be said that all these personal dimensions, and therefore subjective, are precisely those, which should be at the door of the lab for scientific objectivity is not damaged. Furthermore, integration, dosage and balance of all these personal skills are achieved through common sense or, in more philosophical terms thanks to a reasonable attitude.[2]

Personal Action and Scientific Objectivity

It could be argued that science, transformed into a personal matter, inexorably loses objectivity. If emotions are mixed in scientific research, objectivity may suffer. But on the other hand, even without emotions we would get to do science. No one would want to spend his life time, work and effort on a task that will be indifferent, just a personal achievement without any possible objective value. How can we address this apparent paradox?

We can take advantage of a metaphor borrowed from the biological domain. Consider science as the cardio circulatory system: in diastolic phase blood enters into the heart from the whole body, while the contraction or systolic phase leaves a minimal amount of blood in its cavities. Similarly, when we breathe we store in our lungs air from our environment, while the expiration leaves in them a much smaller amount of gas. Science also pulse and breathe, at least metaphorically, at certain stages we find science in need of all the capabilities of the person quoted above, while in other phases must temporarily reduce the presence of some of them and focus primarily on logical inference or observation. Again at certain stages, scientific action requires many external resources, from the social and cultural environment taken from the most diverse traditions, while at other stages these external elements are less present.

We are facing a matter of degree, not black or white. Neither the heart nor the lungs get completely empty when functioning normally and healthy. Similarly, science can never completely ignore the set of personal skills or social environment demands. And of course, the scientific action as a whole is personal action and (therefore) social. These subjective aspects are perfectly compatible with scientific objectivity, which depends from the rate, dosage, harmony and balance of the pulses as a whole, just as the proper functioning of the lungs or heart depends on its rhythms and balances, it could not be achieved only on the basis of single contractions or expansions.

Phases of Scientific Practice

Let us now consider more in detail these pulsatile movements occurring throughout scientific activity. To get started we will distinguish several stages of it. What we are presenting is a simplified diagram of the phases through which it passes scientific activity. It is just one of the possible routes that a scientist may follow, and it is not even intended to represent the actual chronological order of research. In addition, for the sake of brevity, we will put aside all feedback loops. We will not take into account the fractal shape of scientific action: each of the phases in turn consists of sub-phases, and has a certain internal complexity. Each field has its own methodological research idiosyncrasies, with a specific history. Moreover, the scientific activity is not primarily concerned with the mechanical application of any of these methods, but all of them are created at the time of investigation. Suppose, then, with all these caveats, that the scientific activity moves through the following phases:

(i) Identification and problem setting;
(ii) Formulation of hypotheses and selection among them;
(iii) Identification of the auxiliary assumptions and empirical consequences evaluation;
(iv) Observation and experimentation;
(v.i) Predictive empirical verification;
(v.i.i) Explanation and prediction;
(v.i.ii) Transfer and application;
(v.i.iii) Communication and education;
(v.i.iv) Detection or construction and new problems approach;
(v.ii) Provisional empirical falsification;
(v.ii.i) Rethinking the auxiliary hypotheses or assumptions.

The Pulse of Science

Identification and Problems Setting

Now consider the first of the mentioned phase (i), which refers to problems. Scientific research would not even start without it. In this phase problems are identified, sometimes built, for research. It is obvious that the identification of problems depends on our sense of wonder and our curiosity as well as the social environment and the traditions in which we are embedded. Since its inception, then, the scientific activity radically depends on emotions and feelings such as curiosity or amazement. It also depends on social contexts and historical traditions. What was a riddle worthy of investigation to Herodotus (484–425 BC), the cycle of the Nile River, for many others was a simple fact of life that did not request any special explanation. What in a context or tradition can be seen as a problem, may be unnoticed for others. As we saw in this volume, there is still the urge to reflect on the notion of scientific understanding and the arise of the scientific problem (Dieguez, MacLeod): is science creating its own issues? What is a problem for a scientist may not be for another. Why? Maybe it is a difference of aesthetic or social or moral sensibility, maybe a difference of interests. Perhaps some of them are more concerned about consistency, simplicity, and elegance, while others better appreciate the moral, ecological or social nuances, or the practical and functional, or economic, or any personal combination of these or other issues. Sensitivities and motivations are different, conditioned by the personality and circumstances of each person doing science, all of them legitimate, all of them rich when identifying and posing problems, all necessary for collaborative task that is science. A first important suggestion that emerges from history and philosophy of science is that Logos is something wider than mere rationality, or, vice versa, human rationality is wider than Logos reduced to reasoning activity. In its wider meaning, in fact, Logos admits, and not excludes, aesthetical, social and moral attitudes toward life and world.[3]

A person may be attracted or surprised by various research problems. And yet, you probably have to decide, you have to take a course or another. There is not, of course, an algorithm that allows us to make these decisions in a purely mechanical or logical way. All the aspects of the person are required in order to make these decisions. His imagination, intuition, sensitivity, practicality, experience, all of these components will tell you which of the identified problems will deserve your immediate attention, which one seems unapproachable, which may be now postponed and then resumed under better circumstances; his logical sense will tell if any of them depends on the prior resolution of another; the dialogue with colleagues, sometimes even the frequentation of friends as well as the daily experiences and traditional sources of wisdom, can be very helpful to decide. All the aspects of the person and the circumstances he is in play as a coordinated and balanced team thanks to the common sense of the person making science.

At a certain moment we move from problem identification to approach it, in understandable and relevant terms to the scientific community and society in general, with which scientists share a common ground, guaranteed by our common human nature. This transit is likely to require the logical ability to argue, language and rhetorical skills to convince, the power of observation to finish outlining the problem, to confirm, to the possible extent, its relevance and viability. The way we set a problem provides guidance, orientation, connoted from the beginning by a peculiar sensitivity; the target whose service we will put our logical and observational skills will allow us to move from the problem identification to clear and fruitful approach. Of course the need to adopt scientific language will decrease the presence of most emotional, aesthetic or moral aspects, especially in their more idiosyncratic facets, but they will not completely disappear at any time, inter alia, for the person doing science is addressing to people equipped with emotions and interests as much as logic and observation capability.

Hypothesis Formulation and Selection

After the problem statement begins the search of adequate hypothesis to address it (ii). It seems obvious that the production of hypotheses critically depends on the imaginative capacities of the person. Nevertheless, some philosophers have tried to reduce the production of hypotheses to an inference of deductive or inductive nature. Some of them have even thought that this step could be executed in an inferential algorithmic or mechanical way, as we saw in some of the previous contributes, where we had the opportunity to enlighten the notion of computational simulation and modelling in scientific explanation. The modern sensibility tried to elaborate a theory for an universal scientific method. On the one hand, rationalistic philosophers thought that it was possible and therefore necessary to infer scientific laws deductively from certain general or transcendental principles. On the other hand, empiricists argued that the inference principle should be firmly grounded on the observation, so that from repeated observations it is possible to draw certain general laws. In both of these philosophical traditions utterances achieved by means of inference lose their hypothetical characterization, showing themselves since the beginning as certain laws. Nevertheless, we now know that the statements of science never completely lose their hypothetical nature, always with them goes a shadow of uncertainty and tentativeness, however tenuous, and we also know that the production of hypotheses is guided not only by logical inference and observation, but also by whole creative capacity of the individual.

How does this creative act occur? Growing our creativity requires much effort. Though it is possible, we have to admit that we know very little about the ultimate roots of creativity. The French physicist Pierre Duhem (1861–1916) commented wryly that whoever believes that the scientific idea springs from nowhere, as if by magic, he is like the child who sees the chicken coming out of his shell and then thinks it is made at that time, not even imagining the complexity of a long process of gestation. Scientists usually prepare the ground through study, meditation and imaginative exercises, conversation, discussion, observation, reading. Nevertheless, hypothesis, according to Duhem, 'germinates in him, without him'. And, once an idea has been conceived, again his 'free and laborious activity must come into play' to 'develop it and make it fruitful'.[4] We say that our ideas come to mind, not that 'happen to us', what we can do about them, to promote their appearance, is to freely manage conditions under which they may arise. These conditions involve the whole person and depend on the contexts and traditions in which the person is located (inspiratory phase).

The origins of scientific hypothesis must be sought sometimes in the remote places of science, such as the artistic background of a person or his metaphysical or religious beliefs. Consider the historical case of Johannes Kepler (1571–1630).

His struggle of more than a decade to solve the problem of the trajectory of Mars led to the famous hypothesis of the ellipse. Without the accurate observations of Tycho Brahe (1546–1601) this scientific advance would have not occurred and without Kepler's mathematics and logic capacity the discovery would not have been possible.

In fact, both of them knew they needed each other, especially the complementary nature of their skills. But nothing would have arisen, no great new hypothesis, without many other capabilities of Kepler and Brahe, among which are included: determination, discipline, love for scientific work, the admiration professed by the order of heaven, and their immense curiosity. Nothing of interest would have occurred without the knowledge Kepler had in the history of mathematics, and in particular in the older studies on conic curves, without profound metaphysical convictions of their Pythagorean root, without his religious vision of the Cosmos, which for him was a reflection and symbol of the Holy Trinity, without his aesthetic sense and appreciation of simplicity and elegance. The creation of the hypothesis also depended on the personal interests of both the authors, as well as on their social environment, very peculiar, even opposite and complementary in some sense. As we can see, the hypothesis in this case is the result of the whole person and circumstances.

Before moving to the next phase, consider the situation in which a scientist – a person who makes science – or a scientific community must decide among several alternative scenarios in order to address the same problem. It is a common situation. We are not referring to the choice between two or more tested hypotheses, but to the initial choice among emerging hypothesis, a discriminatory work which tells us which hypotheses deserve to be followed and which not. Trying to decide which of them deserve to be empirically tested, which one or ones are most promising. Again we find that such decisions are the result of a personal agency, whereas all the capabilities of the person are orchestrated by common sense. These capabilities such as intuition, experience,[5] contextual and traditional indications, imagination can hardly be formalized.

Auxiliary Assumptions and Empirical Consequences

Once we have considered the problem, generated and selected hypotheses, it is time to consider the phases of empirical verification. To test a hypothesis we need to extract its empirical consequences. But this is impossible if we do not add to the taken hypothesis some assumptions commonly accepted in the body of knowledge (iii). For example, the chemistry developed by Georg Ernest Stahl (1660–1734) tried to explain phenomena such as the calcination of metals, what we now call oxidation, from the hypothesis of phlogiston. According to the same theory, calcination and other phenomena, such as combustion or respiration, always occur together with the release into the atmosphere of a hypothetical substance called phlogiston. It seems to follow, as an empirical result, that the metal will lose weight during the process of calcination. But this result is necessary only if we assume that the weight of phlogiston is positive. As we can see, the empirical conclusion comes from the hypothesis plus a number of accepted assumptions that sometimes overextend themselves from a scientific discipline to other fields. Not to mention the most usual cases, such as those related to the reliability of the instruments, the confidence to the reports from colleagues or our own senses and states of consciousness. What is empirically tested in each case is not an isolated hypothesis but a broad set of assumptions.

We have to choose our auxiliary assumptions and modulate the trust we put in them. Again we are faced with an operation that can only spring from the integrity of the person. Once done, we are in a position to infer from the hypothesis (H) and its auxiliary assumptions (A) any observable and empirical consequence (O). And now the logic is in charge to state: $(H \wedge A) \rightarrow O$. Imagine that Stahl had chosen his hypothesis of phlogiston and a Newtonian concept of mass. It follows that – ceteris paribus – calcination must be accompanied by a weight loss in the metal. Maybe Stahl did not like this empirical result, but his personal taste cannot interfere with the inferential mechanism.

Observation and Experimentation

At this point, we have an empirical statement comparable with observable data (iv). You might think that the empirical phase is completely free of personal items. But we know, at least since the time of Thomas Kuhn (1922–1996), it is not so easy, and even the simplest observation is conditioned by our expectations, to say nothing of complex experiments that are required in many of the scientific disciplines. In many of them, the concrete action of scientists builds the object or phenomenon in question, as revealed Canadian philosopher Ian Hacking (1936). Every perception, even the simplest, is not merely passive, but it is feasible thanks to the activity of the subject. Philosophers have given this phenomenon the name of theory-loading of facts. This suggests that empirical observations are mediated or influenced by the theoretical perspective, or in the words of Kuhn, the paradigm investigation starts from: on one side, Logos determines attitudes in perception of facts (Bios), on the other side, phenomena (Bios) suggests the way of better conceiving them (Logos).

The conditions of observation go far beyond their own theories and scientific paradigms. It is instructive, in this regard, the historical Galileo's (1564–1642)[6] case of lunar observation. As we know, Galileo was the first to use the telescope to observe the heavens. Through it he was able to see some shadows on the moon. Immediately he interpreted these shadows as an evidence of the lunar relief. Everyone knows the famous pictorial representations of the lunar relief made by Galileo.[7] But to see the lunar terrain is not as simple as putting the eye to the telescope, it's not just a better use of Téchne. Proof of this is that around the same time, the British Thomas Harriot (1560–1621) also observed the Moon through a telescope, but from another training and expectations, and he failed to see how relief those patterns of light and shadow that showed to him.[8] In this case, the formation of Galileo, his pictorial techniques of chiaroscuro – his Logos in the above mentioned wider sense – conditioned and facilitated his lunar observations.

These considerations should not lead us to any relativistic conclusion. At the end of the day, the Moon does have relief, and Harriot came to accept it. Luckily, we also have a few sketches of the lunar surface drawn by Harriot. One drawn before and one after seeing the Galileo's description. The two sketches drastically differ. In the first we see a linear cycle divided by a broken line, which represents the boundary of the illuminated spot. In the second clearly appear craters and valleys. Harriot learnt to see the relief of the moon once he read Galileo. Even in the observation phase all personal background is involved. But imagine that Harriot, because of the attachment to his previous positions, interest or nostalgia, had refused to acknowledge what he actually was able to see after reading reports of Galileo. Clearly, when someone has assumed a certain background, what you see is what you see, and other considerations are out of place.

Verification and Falsification

Once obtained the empirical results by observation and experimentation, you can match them, wholly or partly, with theoretical expectations. In case that the match is not perfect, we will appeal to a certain estimation capacity that will tell us if the overlap degree is significant or not. For example, the empirical results obtained by Gregor Mendel (1822–84) in his famous experiments with peas did not conform exactly to the theoretical predictions, but they were very close, enough to give accurate forecasts by taking the slight deviations as mere disturbances.

Galileo suggested something similar in his *Discorsi*, when he admits that the empirical data on the drop of subjects to gravity do not match exactly with his theoretical predictions, but Aristotelian physics brings much greater imbalances that can no longer be attributed to disturbances. The respective uncertainties are orders of magnitude distant; they differ, as Galileo says, as a grain of sand from a millstone, as a hair from a rope.

Apparently, both Mendel and Galileo were justified to ignore small inaccuracies attributable to uncontrolled disturbances. The empirical results, in spite of them, confirmed their hypothesis. It is not easy to decide when a divergence of this type is negligible and when it is not. Such estimations depend – again – on numerous capacities and circumstances ruled by the wisdom of the person doing science. But once the decision to take or not take into account the discrepancy has been made, we find empirical evidence that either reinforces or refutes the hypothesis. In the first instance we have verification (vi), in the second falsification (v.ii). Now, we must ask to what extent an empirical success or failure can make an hypothesis respectively true or false. The logical scheme of verification would be this:

$$(H \wedge A) \to O$$
$$O$$

From these two premises cannot be legitimately inferred the assertion of the antecedent of the first one ($H \wedge A$), pure logic does not allow us to assume as verified a hypothesis after its empirical confirmation. Neither the accumulation of empirical successes helps. It is of no help even to grow the probability of the hypothesis, since the probability is calculated by the ratio of favorable cases on possible ones. When possible cases are infinite, as it is often the case, the accumulation of favorable cases does not grow the probability. And yet, something tells us that empirical success should have some meaning for the corresponding hypothesis. And it does, but the way it does cannot be grasped by pure logic, neither from the calculus of probability, this meaning comes precisely from the set of human faculties that a person can handle harmoniously through common sense.

The same is true regarding falsification, whose formal scheme goes as follows:

$$(H \wedge A) \to O$$
$$\neg O$$

$$\neg(H \wedge A)$$

$$\neg H \vee \neg A$$

I.e., when the observed $\neg O$, diverges significantly from the expected O, we know that something is wrong in the premise, meaning that $\neg(H \wedge A)$, but we cannot know from pure logic if what troubles are the hypothesis or any of the auxiliary assumptions, i.e. we just know that either $\neg H$ or $\neg A$. What should we do in this case? Do we reject the hypothesis or any of the underlying assumptions?

There are plenty of historical cases for all tastes. Here we will put a couple of examples so you can have a glimpse to the complexity of the case. Some philosophers have drawn, perhaps guided by their psychological preferences, interests, feelings or metaphysical belief, consequences from empirical data that logic simply does not allow. Thus verificationists, as Rudolf Carnap (1891–1970), pretended empirical success to be considered as definitive, secure and permanent verification of a hypothesis. In a weaker version, probabilistic, it was claimed that the succession of empirical successes serves to generate an increased probability of the hypothesis. Meanwhile, falsificationists as Karl Popper argued that the refutation of a hypothesis is empirically rooted. But logic does not allow such conclusions, and logic tells us that even after several observation, the decisions are open: we can choose to keep or discard our hypothesis.

Now let's move to the historical cases. Consider again the case of phlogiston. Given the observation that metals are calcined with weight gain, the phlogiston theory can choose to discard the hypothesis or keeping it at the cost of reviewing any of the auxiliary assumptions. You can propose, for example, that phlogiston actually has a negative weight. It is clear that this move will put to new difficulties, but it is a possibility. With the advantage that gives us the history we would say, 'give up your hypothesis, phlogiston does not exist'.

Well, what would we say to Copernicus? According to his hypothesis, the Earth orbits around the Sun, that is, that we as terrestrial observers toured the area, and see the stars from a position that changes throughout the year. Hence it follows an empirical prediction: we should see how the shape of the constellations will change throughout the year. Stated more technically, we should observe stellar parallax. Copernicus looks at the sky and did not see the parallax. The constellations keep their shape throughout the year. Then either the hypothesis is false, or have accepted an incorrect premise. Standing on shoulders of giants, we say to Copernicus the opposite of what you would have told to Stahl: 'resist, heed your intuition, your aesthetic sense of harmony, your Sun Pythagoreanism, do not give your hypothesis up, the Earth itself orbits, check the auxiliary assumptions'. That was what made Copernicus. He advocated reviewing a commonly accepted assumption on the size of the universe. He suggested that perhaps the universe is much larger than it had been assumed so far. Thus, the stars would be so far from us that the size of Earth's orbit would be negligible in relation to the distance that separates us from them. He did not live long enough to look through a good telescope. Had he got the chance to take a glance into it, indeed, he could have seen the stellar parallax his hypothesis predicted and which could have not be seen by naked eye, given the enormous distance that separates us from even the nearest star (Alpha Centauri).

What is clear is that we play with advantage, though we don't have an algorithm or an uniform method that allows us to recommend, at a certain observation, the way to go. We would recommend to Stahl the opposite of Copernicus. We already know that the scientific action is personal action, but there is no universal recipe to tell us at what extent we should combine all facets of the person, in which phases should be more restrictive or more inclusive. The same pictorial formation gave an advantage to Galileo respect to Harriot, but prevented him from accepting the elliptical orbits proposed by Kepler. Galileo rejected them because of his preference for classical circle and his contempt for the ellipsoids forms of Mannerist. However, we want the investigation to be rational. But here rationality is understood as common sense, not as an algorithm. There is rationality where there are rational beings, i.e. people. As Pierre Duhem said, the message of science is directed to common sense (bon sens).

Explanation, Application and More

Always from common sense we decide what assumptions need to be accepted and with what degree of confidence. Hence – via deductive inference – we look for explanations and predictions that can be made (vii). Take this time a contemporary case. In the field of oncology there are currently two main hypotheses in contention, each with its variants. First one is the SMT (Somatic Mutation Theory), whereby genetic mutations in somatic cells are considered the primary cause of cancerous growth, the second one is TOFT (Tissue Organization Field Theory), whereby cancer proceeds, ultimately, because of incorrect tissue organization. Each hypothesis provides an alternative explanation of the diseases in question. Here, again, come into play other considerations beyond inference and observation. In the foreground is the success of each of the hypotheses in explaining phenomena linked to cancer, such as carcinogenesis, metastasis, cell heterogeneity, and other such spontaneous reversion.

Karl Popper used to say that science progresses from problems to problems. If we reject a hypothesis we have to rethink the original problem as well (v.ii.i), and if we accept it a whole new kind of problem (v.i.iv) appears. That is, the explanation of a phenomenon using a hypothesis is not only one end of the route, but also the beginning of the identification and setting of new problems, may be deeper, or more accurate, or better formulated in which the accepted hypotheses serve as a heuristic guide. If we adopt the SMT, we are trying to identify oncogenes, whereas if we accept the TOFT,[9] we are orienting the search towards pathogenic tissue configurations.

Another common result of the acceptance of a hypothesis is the transfer and application (viii) or the attempt of application of it in the practical and technological field. Here we move in a domain of great complexity, as there is almost never a possible automatic application of a hypothesis to solve practical problems, every application instead requires a certain art, a certain adaptability to terrain. It is not possible here to present the complexity of applied science and technology. It is enough to our purpose to remember that, indeed, each of these levels is in itself an art and a complex tradition, and not the result of a simple mechanical translation of theoretical science. Even the phase called scientific transfer, which passes knowledge gained from research to the production system, requires wisdom itself. Both the transfer and implementation require new balanced contribution of a large number of human capabilities. Consider the difficulties involved in transferring to the pharmaceutical industry scientific knowledge about cancer, so that it is possible to derive new applications. Imagine, for a moment, the amount of human and social skills required before being able to see an effective medication at the pharmacy, and how delicate it can be to work balancing them all.

The same is true for communication and education (v.i.iii). The science communication is done through the media, the popular works of science museums, film and literature, among other channels. It is obvious that this is not an automatable activity and it is not enough to have scientific knowledge to run successfully this system. It is a task of mediation in which the communicator has to create new metaphors, he has to perform certain critical freedom as well, and must find a new balance between different values such as rigor and amenity, for example.[10] Also the teaching of science in schools and universities, requires that all human capacities, implied in both teaching and studying, work harmoniously, from emotional intelligence to the social one, from creativity to empathy, from expository clarity to intellectual honesty and surely many others. It follows that the teaching of science does not pursue the training of scientists, but focuses on people capable of doing science.

Conclusion

The aim of this chapter is to present science as an activity performed by people, with all their skills, attitudes and circumstances. Science is not a modular activity, which can be reduced to the powers of observation and logical inference of a single researcher. It rather is an integral personal activity. Conforming to this idea, we presented scientific rationality as a kind of harmony or equilibrium, as the result of the dosage and timing where all contextual circumstances and capabilities are combined. We have adopted here the metaphors of heartbeat and respiration to suggest precisely these images and dose rate. According to the metaphor of the heartbeat, the person doing science takes, depending on the moment, a higher or lower dose of each of its capacities. According to the metaphor of breath, this person also includes in his scientific work more or less dose of social or historical circumstances of their environment. The dosage and the rate charged to the common sense of the person, his wisdom or prudence.

Harmony and equilibrium between different dimensions of human person shed light on the mutual relationship between Bio-Techno-Logos in scientific practice. There can't be a primate of Bios, Techno nor Logos, otherwise all the scientific practice fall under a one-sided approach. On left side of the triad, phenomena (Bios) are strictly related to instruments adopted (Téchne) by scientists and also to the theory (Logos) in which they are explained. On the opposite side, Logos is something you never completely understand in its complexity and multidimensionality. We can speak about Logos in scientific terms, in logical terms, in aesthetical terms, in ethical terms, in spiritual terms. There is no privileged perspective that grasps the vitality and fertility of human Logos. Even in the restricted area of scientific research, no privileged dimension of Logos has to be evoked as the most important one. Surely, logical or inferential dimension may result decisive, but not the only one. The traditional alternative between context of the discovery and context of the justification split up Logos into different areas and personal attitudes supposed to play different roles. On the contrary, in scientific practice, discover and justification are the expression of the same rational inquiring activity on world and are thus the expression of the human person as a whole. Last, as we learned from history of science, Téchne is never just Téchne, since it implies a Bios to be inquired and a Logos able to plan the adequate way to inquire. Therefore, Téchne is fully imbued with both Logos and Bios: that's the reason why sometimes it is conceived as raw material, like technological instruments or robotic environment, and some other it is a refined product of pure reason, as analysis situs or integral calculus.

Unity, complexity and harmony of the person do reflect in scientific practice where the principal actors are humans, not theories nor supposed natural processes. Reducing scientific research to search for truth, to human enhancement, to the elaboration of more refined theories, or to a better understanding of specific topics, forgets the main basilar element of every scientific practice that is the fact that it always remains a *human* attitude. So, as a human practice, science reveals its deepest meaning, which has to be anthropological rather than logical or merely methodological.

NOTES

1 Selvarajoo, Microscopic and Macroscopic Insights on Dynamic Cell Behaviour

1. R. Nilsson, V. B. Bajic, H. Suzuki, D. di Bernardo, J. Björkegren, S. Katayama, J. F. Reid, M. J. Sweet, M. Gariboldi, P. Carninci, Y. Hayashizaki, D. A. Hume, J. Tegner and T. Ravasi, 'Transcriptional Network Dynamics in Macrophage Activation', *Genomics*, 88 (2006), pp. 133–42.
2. H. Jeong, S. P. Mason, A. L. Barabási and Z. N. Oltvai, 'Lethality and Centrality in Protein Networks', *Nature*, 411 (2001), pp. 41–2.
3. M. Vidal, M. E. Cusick and A. L. Barabási, 'Interactome Networks and Human Disease', *Cell*, 144 (2011), pp. 986–98.
4. Jeong, Mason et al., 'Lethality and Centrality in Protein Networks', pp. 41–2; Vidal, Cusick and Barabási, 'Interactome Networks and Human Disease', pp. 986–98.
5. A. Naito, S. Azuma, S. Tanaka, T. Miyazaki, S. Takaki, K. Takatsu, K. Nakao, K. Nakamura, M. Katsuki, T. Yamamoto and J. Inoue, 'Severe Osteopetrosis, Defective Interleukin-1 Signalling and Lymph Node Organogenesis in TRAF6-Deficient Mice', *Genes Cells*, 4 (1999), pp. 353–62; T. Kobayashi, M. C. Walsh and Y. Choi, 'The Role of TRAF6 in Signal Transduction and the Immune Response', *Microbes and Infection*, 6 (2004), pp. 1333–8.
6. L. Dartnell, E. Simeonidis, M. Hubank, S. Tsoka, I. D. Bogle and L. G. Papageorgiou, 'Robustness of the p53 Network and Biological Hackers', *FEBS Letters*, 579 (2005), pp. 3037–42.
7. P. E. Purnick and R. Weiss, 'The Second Wave of Synthetic Biology: From Modules to Systems', *Nature Reviews Molecular Cell Biology*, 10 (2009), pp. 410–22.
8. S. L. Werner, D. Barken and A. Hoffmann, 'Stimulus Specificity of Gene Expression Programs Determined by Temporal Control of IKK Activity', *Science*, 309 (2005), pp. 1857–61; R. Cheong, A. Hoffmann and A. Levchenko, 'Understanding NF-kappaB Signaling via Mathematical Modeling', *Molecular Systems Biology*, 4 (2008), ep. 192.
9. S. D. Santos, P. J. Verveer and P. I. Bastiaens, 'Growth Factor-Induced MAPK Network Topology Shapes Erk Response Determining PC-12 Cell Fate', *Nature Cell Biology*, 9 (2007), pp. 324–30.
10. F. J. Bruggeman, H. V. Westerhoff, J. B. Hoek and B. N. Kholodenko, 'Modular Response Analysis of Cellular Regulatory Networks', *Journal of Theoretical Biology*, 218 (2002), pp. 507–20.

11. R. H. Blair, D. L. Trichler and D. P. Gaille, 'Mathematical and Statistical Modeling in Cancer Systems Biology', *Frontiers in Physiology*, 3:227 (2012), pp. 1–8; K. Selvarajoo, M. Tomita and M. Tsuchiya, 'Can Complex Cellular Processes be Governed by Simple Linear Rules?', *Journal of Bioinformatics and Computational Biology*, 7 (2009), pp. 243–68.

12. A. Voss-Böhme, 'Multi-Scale Modeling in Morphogenesis: A Critical Analysis of the Cellular Potts Model', *PLoS One*, 7 (2012), ep. 42852; Y. Yamada and D. Forger, 'Multi-scale Complexity in the Mammalian Circadian Clock', *Current Opinion in Genetics and Development*, 20 (2010), pp. 626–33.

13. M. Strasser, F. J. Theis and C. Marr, 'Stability and Multiattractor Dynamics of a Toggle Switch Based on a Two-Stage Model of Stochastic Gene Expression', *Biophysical Journal*, 102 (2012), pp. 19–29; D. Stockholm, F. Edom-Vovard, S. Coutant, P. Sanatine, Y. Yamagata, G. Corre, L. Le Guillou, T. M. Neildez-Nguyen and A. Pàldi, 'Bistable Cell Fate Specification as a Result of Stochastic Fluctuations and Collective Spatial Cell Behaviour', *PLoS One*, 5 (2010), ep. 14441.

14. G. J. Baart and D. E. Martens, 'Genome-Scale Metabolic Models: Reconstruction and Analysis', *Methods in Molecular Biology*, 799 (2012), pp. 107–26; C. H. Schilling and B. O. Palsson, 'The Underlying Pathway Structure of Biochemical Reaction Networks', *Proceedings of the National Academy of Sciences of the United States of America*, 95 (1998), pp. 4193–8.

15. M. K. S. Yeung, J. Tegner and J. J. Collins, 'Reverse Engineering Gene Networks Using Singular Value Decomposition and Robust Regression', *Proceedings of the National Academy of Sciences of the United States of America*, 99 (2002), pp. 6163–8; O. Alter, P. O. Brown and D. Botstein, 'Singular Value Decomposition for Genome-Wide Expression Data Processing and Modeling', *Proceedings of the National Academy of Sciences of the United States of America*, 97 (2000), pp. 10101–6.

16. N. Friedman, M. Linial, I. Nachman and D. Pe'er, 'Using Bayesian Network to Analyze Expression Data', *Journal of Computational Biology*, 7 (2000), pp. 601–20.; N. Friedman, 'Inferring Cellular Networks Using Probabilistic Graphical Models', *Science*, 303 (2004), pp. 799–805; Z. X. Yeo, S. T. Wong, S. N. Arjunan, V. Piras, M. Tomita, K. Selvarajoo, A. Giuliani and M. Tsuchiya, 'Sequential Logic Model Deciphers Dynamic Transcriptional Control of Gene Expressions', *PLoS One*, 2 (2007), ep. 776.

17. K. Selvarajoo and M. Tomita, 'Physical Laws Shape Biology', *Science*, 339 (2013), p. 646.

18. K. Selvarajoo, Y. Takada, J. Gohda, M. Helmy, S. Akira, M. Tomita, M. Tsuchiya, J. Inoue and K. Matsuo, 'Signaling Flux Redistribution at Toll-Like Receptor Pathway Junctions', *PLoS One*, 3 (2008) ep. 3430; W. Vance, A. Arkin and J. Ross, 'Determination of Causal Connectivities of Species in Reaction Networks', *Proceedings of the National Academy of Sciences of the United States of America*, 99 (2002), pp. 5816–21; M. Helmy, J. Gohda, J. Inoue, M. Tomita, M. Tsuchiya and K. Selvarajoo, 'Predicting Novel Features of Toll-Like Receptor 3 Signaling in Macrophages', *PLoS One*, 4 (2009), ep. 4661.

19. Selvarajoo, Tomita and Tsuchiya, 'Can Complex Cellular Processes be Governed by Simple Linear Rules?', pp. 243–68.

20. Selvarajoo, Takada et al., 'Signaling Flux Redistribution at Toll-Like Receptor Pathway Junctions', ep. 3430; K. Selvarajoo, 'Discovering Differential Activation Machinery of the Toll-Like Receptor 4 Signaling Pathways in MyD88 Knockouts', *FEBS Letters*, 580 (2006), pp. 1457–64.

21. K. Selvarajoo, 'Decoding the Signalling Mechanism of Toll-Like Receptor 4 Signaling Pathways in Wildtypes and Knockouts', in S. Arjunan, P. K. Dhar and M. Tomita (eds), *E-Cell System: Basic Concepts and Applications* (New York and Austin, TX: Springer and Landes Bioscience, 2013), pp. 157–67.

22. Selvarajoo, Takada et al., 'Signaling Flux Redistribution at Toll-Like Receptor Pathway Junctions', ep. 3430; Selvarajoo, Tomita and Tsuchiya, 'Can Complex Cellular Processes be Governed by Simple Linear Rules?', pp. 243–68.

23. Selvarajoo, Takada et al., 'Signaling Flux Redistribution at Toll-Like Receptor Pathway Junctions', ep. 3430; J. C. Kagan, T. Su, T. Horng, A. Chow, S. Akira and R. Medzhitov, 'TRAM Couples Endocytosis of Toll-Like Receptor 4 to the Induction of Interferon-Beta', *Nature Immunology*, 9 (2008), pp. 361–8; I. Zanoni, R. Ostuni, L. R. Marek, S. Barresi, R. Barbalat, G. M. Barton, F. Granucci and J. C. Kagan, 'CD14 Controls the LPS-Induced Endocytosis of Toll-Like Receptor 4', *Cell*, 147 (2011), pp. 868–80.

24. V. Piras, K. Hayashi, M. Tomita and K. Selvarajoo, 'Enhancing Apoptosis in TRAIL-Resistant Cancer Cells Using Fundamental Response Rules', *Scientific Reports*, 1 (2011), ep.144.

25. K. Hayashi, V. Piras, S. Tabata, M. Tomita and K. Selvarajoo, 'A Systems Biology Approach to Suppress TNF-Induced Proinflammatory Gene Expressions', *Cell Communication and Signaling*, 11 (2013), ep. 84.

26. M. B. Elowitz, A. J. Levine, E. D. Siggia and P. S. Swain, 'Stochastic Gene Expression in a Single Cell', *Science*, 297 (2002), pp. 1183–6.

27. J. M. Raser and E. K. O'Shea, 'Control of Stochasticity in Eukaryotic Gene Expression', *Science*, 304 (2004), pp. 1811–14; W. J. Blake, M. Kaern, C. R. Cantor and J. J. Collins, 'Noise in Eukaryotic Gene Expression', *Nature*, 422 (2003), pp. 633–7.

28. H. H. Chang, M. Hemberg, M. Barahona, D. E. Ingber and S. Huang, 'Transcriptome-Wide Noise Controls Lineage Choice in Mammalian Progenitor Cells', *Nature*, 453 (2008), pp. 544–7.

29. E. Lorenz, *The Essence of Chaos* (Seattle, WA: University of Washington Press, 1995).

30. H. Maamar, A. Raj, and D. Dubnau, 'Noise in gene expression determines cell fate in Bacillus subtilis', *Science*, 317 (2007) pp. 526–9.

31. G. M. Süel, J. Garcia-Ojalvo, L. M. Liberman and M. B. Elowitz, 'An Excitable Gene Regulatory Circuit Induces Transient Cellular Differentiation', *Nature*, 440 (2006), pp. 545–50.

32. K. Selvarajoo, 'Understanding Multimodal Biological Decisions from Single Cell and Population Dynamics', *Wiley Interdiscip Rev Syst Biol Med*, 4 (2012), pp. 385–99.

33. J. Jullien, V. Pasque, R. P. Halley-Stott, K. Miyamoto and J. B. Gurdon, 'Mechanisms of Nuclear Reprogramming by Eggs and Oocytes: A Deterministic Process?', *Nature Reviews Molecular Cell Biology*, 12 (2011), pp. 453–9.

34. Z. D. Smith, I. Nachman, A. Regev and A. Meissner, 'Dynamic Single-Cell Imaging of Direct Reprogramming Reveals an Early Specifying Event', *Nature Biotechnology*, 28 (2010), pp. 521–6.

35. E. Lorenz, *The Essence of Chaos* (Seattle, WA: University of Washington Press, 1995).

36. E. Lorenz, 'Deterministic Nonperiodic Flow', *Journal of the Atmospheric Sciences*, 20 (1963), pp. 130–48.

37. L. S. Liebovitch and D. Scheurle, 'Two Lessons from Fractals and Chaos', *Complexity*, 5 (2000), pp. 34–43.

38. H. Poincaré, 'Les méthodes nouvelles de la méchanique céleste', vol. III (Paris: Gauthiers, 1899).

39. Liebovitch and Scheurle, 'Two Lessons from Fractals and Chaos'.
40. S. Huang, G. Eichler, Y. Bar-Yam and D. E. Ingber, 'Cell Fates as High-Dimensional Attractor States of a Complex Gene Regulatory Network', *Phys Rev Lett.*, 94 (2005), ep. 128701; M. Tsuchiya, V. Piras, A. Giuiliani, M. Tomita and K. Selvarajoo, 'Collective Dynamics of Specific Gene Ensembles Crucial for Neutrophil Differentiation: The Existence of Genome Vehicles Revealed', *PLoS One*, 5 (2010), ep. 12116.
41. H. H. Chang, M. Hemberg, M. Barahona, D. E. Ingber and S. Huang, 'Transcriptome-Wide Noise Controls Lineage Choice in Mammalian Progenitor Cells', *Nature*, 453 (2008), pp. 544–7.
42. P. B. Gupta, C. M. Fillmore, G. Jiang, S. D. Shapira, K. Tao, C. Kuperwasser and E. S. Lander, 'Stochastic State Transitions Give Rise to Phenotypic Equilibrium in Populations of Cancer Cells', *Cell*, 146 (2011), pp. 633–44.
43. S. Huang, I. Ernberg and S. Kauffman, 'Cancer Attractors: A Systems View of Tumors from a Gene Network Dynamics and Developmental Perspective', *Seminars in Cell and Developmental Biology*, 20 (2009), pp. 869–76; Y. Guo, G. S. Eichler, Y. Feng, D. E. Ingber and S. Huang, 'Towards a Holistic, Yet Gene-Centered Analysis of Gene Expression Profiles: A Case Study of Human Lung Cancers', *Journal of Biomedicine and Biotechnology*, 2006 (2006), ep. 69141.
44. M. Tsuchiya, V. Piras, S. Choi, S. Akira, M. Tomita, A. Giuliani and K. Selvarajoo, 'Emergent Genome-Wide Control in Wildtype and Genetically Mutated Lipopolysaccarides-Stimulated Macrophages', *PLoS One*, 4 (2009), ep. 4905; M. Tsuchiya, K. Selvarajoo, V. Piras, M. Tomita and A. Giuliani, 'Local and Global Responses in Complex Gene Regulation Networks', *Physica A*, 388 (2009), pp. 1738–46.
45. R. Benzi, 'Stochastic Resonance in Complex Systems', *Journal of Statistical Mechanics: Theory and Experiment*, 1 (2009), ep. 1052.
46. A. Longtin, 'Stochastic Resonance in Neuron Models', *Journal of Statistical Physics*, 70 (1993), pp. 309–27.
47. A. Hodgkin and A. Huxley, 'A Quantitative Description of Membrane Current and its Application to Conduction and Excitation in Nerve', *Journal of Physiology*, 117 (1952), pp. 500–44.
48. N. Suzuki, C. Furusawa and K. Kaneko, 'Oscillatory Protein Expression Dynamics Endows Stem Cells with Robust Differentiation Potential', *PLoS One*, 6 (2011), ep. 27232.
49. A. D. Lander, K. K. Gokoffski, F. Y. Wan, Q. Nie and A. L. Calof, 'Cell Lineage and the Logic of Proliferative Control', *PLoS Biology*, 7 (2009), ep. 1000015; D. C. Kirouac, G. J. Madlambayan, M. Yu, E. A. Sykes, C. Ito and P. W. Zandstra, 'Cell–Cell Interaction Networks Regulate Blood Stem and Progenitor Cell Fate', *Molecular Systems Biology*, 5 (2009), ep. 293.
50. H. H. McAdams and A. Arkin, 'Stochastic Mechanisms in Gene Expression', *Proceedings of the National Academy of Sciences of the United States of America*, 94 (1997), pp. 814–9; C. V. Rao, D. M. Wolf and A. P. Arkin, 'Control, Exploitation and Tolerance of Intracellular Noise', *Nature*, 420 (2002), pp. 231–7; J. P. Capp, 'Stochastic Gene Expression, Disruption of Tissue Averaging Effects and Cancer as a Disease of Development', *Bioessays*, 27 (2005), pp. 1277–85; G. Hornung and N. Barkai, 'Noise Propagation and Signaling Sensitivity in Biological Networks: A Role for Positive Feedback', *PLoS Computational Biology*, 4 (2008), ep. 8.
51. H. Maamar, A. Raj and D. Dubnau, 'Noise in Gene Expression Determines Cell Fate in Bacillus subtilis', *Science*, 317 (2007), pp. 526–9.
52. C. Ribrault, K. Sekimoto and A. Triller, 'From the Stochasticity of Molecular Processes

to the Variability of Synaptic Transmission', *Nature Reviews Neuroscience*, 12 (2011), pp. 375–87; H. H. Lee, M. N. Molla, C. R. Cantor and J. J. Collins, 'Bacterial Charity Work Leads to Population-Wide Resistance', *Nature*, 467 (2010), pp. 82–5.

53. U. Alon, M. G. Surette, N. Barkai and S. Leibler, 'Robustness in Bacterial Chemotaxis', *Nature*, 397 (1999), pp. 168–71.

54. T. L. To and N. Maheshri, 'Noise can Induce Bimodality in Positive Transcriptional Feedback Loops without Bistability', *Science*, 327 (2010), pp. 1142–5.

55. Ribrault, Sekimoto and Triller, 'From the Stochasticity of Molecular Processes to the Variability of Synaptic Transmission'.

56. To and Maheshri, 'Noise can Induce Bimodality in Positive Transcriptional Feedback Loops without Bistability'.

57. K. Selvarajoo, 'Macroscopic Law of Conservation Revealed in the Population Dynamics of Toll-Like Receptor Signaling', *Cell Communication and Signaling*, 9 (2011), ep. 9; V. Piras, M. Tomita and K. Selvarajoo, 'Is Central Dogma a Global Property of Cellular Information Flow?', *Frontiers in Physiology*, 3 (2012), ep. 439.

2 Giuliani, News from the 'Twilight Zone': Protein Molecules between the Crystal and the Watch

1. A. Giuliani, R. Benigni, J. P. Zbilut, C. L. Webber, P. Sirabella and A. Colosimo, 'Nonlinear Signal Analysis Methods in the Elucidation of Protein Sequence/Structure Relationships', *Chemical Reviews*, 102 (2002), pp. 1471–91; W. R. Taylor, 'Protein Structure Comparison Using Bipartite Graph Matching and its Application to Protein Structure Classification', *Molecular & Cellular Proteomics*, 1:4 (2002), pp. 334–9.

2. C. Brandon and J. Tooze, *Introduction to Protein Structure* (New York: Garland, 1999).

3. Giuliani, Benigni et al., 'Nonlinear Signal Analysis Methods in the Elucidation of Protein Sequence/Structure Relationships'.

4. A. K. Dunker, I. Silman, V. N. Uversky and J. L. Sussman, 'Function and Structure of Inherently Disordered Proteins', *Current Opinion in Structural Biology*, 18:6 (2008), pp. 756–64.

5. B. Alberts, D. Bray, J. Lewis, M. Raff, K. Roberts and J. D. Watson, *Molecular Biology of the Cell* (New York: Garland, 1999).

6. B. Modrek and C. Lee, 'A Genomic View of Alternative Splicing', *Nature Genetics*, 30:1 (2002), pp. 13–19; M. Mann and O. N. Jensen, 'Proteomic Analysis of Post-Translational Modifications', *Nature Biotechnology*, 21:3 (2003), pp. 255–61.

7. Modrek and Lee, 'A Genomic View of Alternative Splicing', pp. 13–19; Mann and Jensen, 'Proteomic Analysis of Post-Translational Modifications', pp. 255–61.

8. Giuliani, Benigni et al., 'Nonlinear Signal Analysis Methods in the Elucidation of Protein Sequence/Structure Relationships', pp. 1471–91.

9. L. Di Paola, P. Paci, D. Santoni, M. De Ruvo and A. Giuliani, 'Proteins as Sponges: A Statistical Journey along Protein Structure Organization Principles', *J. Chem. Inf. Model*, 52:2 (2012), pp. 474–82.

10. G. Careri, A. Giansanti and J. A. Rupley, 'Critical Exponents of Protonic Percolation in Hydrated lysozyme Powders', *Physical Review A*, 37:7 (1988), pp. 2703.

11. Careri, Giansanti and Rupley, 'Critical Exponents of Protonic Percolation in Hydrated lysozyme Powders', pp. 2703; J. Kysilka and J. Vondrášek, 'A Systematic Method for Analysing the ProteinHydration Structure of T4 Lysozyme', *Journal of Molecular*

Recognition, 26:10 (2013), pp. 479–87.

12. M. J. Denton, C. J. Marshall and M. Legge, 'The Protein Folds as Platonic Forms: New Support for the Pre-Darwinian Conception of Evolutionby Natural Law', *Journal of Theoretical Biology*, 219:3 (2002), pp. 325–42.

13. Brandon and Tooze, *Introduction to Protein Structure*; Denton, Marshall, and Legge, 'The Protein Folds as Platonic Forms: New Support for the Pre-Darwinian Conception of Evolutionby Natural Law', pp. 325–42.

14. R. Zwanzig, A. Szabo and B. Bagchi, 'Levinthal's Paradox', *Proc. Natl. Acad. Sci*, 89 (1992), pp. 20–2.

15. Denton, Marshall and Legge, 'The ProteinFolds as Platonic Forms: New Support for the Pre-Darwinian Conception of Evolutionby Natural Law', pp. 325–42.

16. H. Frauenfelder and P. Wolynes, 'Biomolecules: Where Physics of Simplicity and Complexity Meet', *Phys. Today*, 47 (1994), pp. 58–61.

17. M. Dokarry, C. Laurendon and P. E. O'Maille, 'Automating Gene Library Synthesis by Structure-Based Combinatorial Protein Engineering: Examples from Plant Sesquiterpene Synthases', *Methods in Enzymology*, 515 (2011), pp. 21–42.

18. P. Pecina, H. Houstková, H. Hansíková, J. Zeman and J. Houstek, 'Genetic Defects of Cytochrome C Oxidase Assembly', *Physiol.Res.*, 53: Suppl 1 (2004), pp. S213–23.

19. C. S. Goh and F. Cohen, 'Co-Evolutionary Analysis Reveals Insight into Protein–Protein Interactions', *J.Mol.Biol*, 324 (2002), pp. 177–92; A. K. Ramani and E. M. Marcotte, 'Exploiting Co-Evolution of Interacting Proteins to Discover Interaction Specificity', *J.Mol.Biol.*, 327 (2003), pp. 273–84.

20. Goh and Cohen, 'Co-Evolutionary Analysis Reveals Insight into Protein–Protein Interactions', pp. 177–92.

21. A. Giuliani, R. Bruni, M. Ciccozzi, A. Lo Presti, M. Equestre, C. Marcantonio and A. R. Ciccaglione, 'Amino-Acid Correlated Mutations Inside a Single Protein System: A New Method for the Identification of Main Coherent Directions of Evolutive Changes', *J. Phylogen. Evolution. Biol.*, 1 (2013), p. 111.

22. Denton, Marshall and Legge, 'The Protein Folds as Platonic Forms: New Support for the Pre-Darwinian Conception of Evolution by Natural Law', pp. 325–42; S. A. Kauffman, *The Origins of Order* (New York: Oxford University Press, 1993); S. Kauffman, *At Home in the Universe: The Search for the Laws of Self-Organization and Complexity* (New York: Oxford University Press, 1995).

23. See http://www.rcsb.org/pdb/home/home.do [accessed February 2015].

24. Denton, Marshall and Legge, 'The Protein Folds as Platonic Forms: New Support for the Pre-Darwinian Conception of Evolutionby Natural Law', pp. 325–42.

25. Denton, Marshall and Legge, 'The Protein Folds as Platonic Forms: New Support for the Pre-Darwinian Conception of Evolution by Natural Law', pp. 325–42.

26. L. Di Paola, M. De Ruvo, P. Paci, D. Santoni and A. Giuliani, 'Protein Contact Networks: An Emerging Paradigm in Chemistry', *Chemical Reviews*, 113 (2013), pp. 1598–613.

27. Di Paola, De Ruvo et al., 'Protein Contact Networks: An Emerging Paradigm in Chemistry', pp. 1598–613.

28. S. Tasdighian, L. Di Paola, M. De Ruvo, P. Paci, D. Santoni, P. Palombo, G. Mei, A. Di Venere and A. Giuliani, 'Modules Identification in Protein Structures: The Topological and Geometrical Solution', *J. Chem. Inf. and Model.*, 54 (2014), pp. 159–68.

29. Di Paola, De Ruvo et al., 'Protein Contact Networks: An Emerging Paradigm in Chemistry', pp. 1598–613.

30. Denton, Marshall and Legge, 'The Protein Folds as Platonic Forms: New Support for the Pre-Darwinian Conception of Evolution by Natural Law', pp. 325–42.
31. S. Kauffman, *At Home in the Universe: The Search for the Laws of Self-Organization and Complexity* (New York: Oxford University Press, 1995).

3 Hang, Limits to Deterministic-Linear Causality in Biomedicine: Effects of Stochasticity and Non-Linearity in Molecular Networks

1. M. Pigliucci, 'Genotype-Phenotype Mapping and the End of the "Genes as Blueprint" Metaphor', *Philos Trans R Soc Lond B Biol Sci*, 365:1540 (2010), pp. 557–66.
2. S. Huang and S. Kauffman, 'How to Escape the Cancer Attractor: Rationale and Limitations of Multi-Target Drugs', *Semin Cancer Biol*, 23:4 (2013), pp. 270–8.
3. S. Huang and J. Wikswo, 'Dimensions of Systems Biology', *Rev Physiol Biochem Pharmacol*, 157 (2006), pp. 81–104.
4. Huang and Wikswo, 'Dimensions of Systems Biology', pp. 81–104.
5. E. Mayr, 'Cause and Effect in Biology', *Science*, 134 (1961), pp. 1501–6; N. Tinbergen, 'Derived Activities; Their Causation, Biological Significance, Origin, and Emancipation during Evolution', *Q Rev Biol*, 27:1 (1952), pp. 1–32.
6. M. Kaern, T. C. Elston, W. J. Blake and J. J. Collins, 'Stochasticity in Gene Expression: From Theories to Phenotypes', *Nat Rev Genet*, 6:6 (2005), pp. 451–64; A. Raj and A. van Oudenaarden, 'Nature, Nurture, or Chance: Stochastic Gene Expression and its Consequences', *Cell*, 135:2 (2008), pp. 216–26.
7. N. Tinbergen, 'Derived Activities; their Causation, Biological Significance, Origin, and Emancipation during Evolution', pp. 1–32; Raj and van Oudenaarden, 'Nature, Nurture, or Chance: Stochastic Gene Expression and its Consequences', pp. 216–26.
8. J. J. Kupiec, 'A Probabilist Theory for Cell Differentiation, Embryonic Mortality and DNA C-Value Paradox', *Speculation Scie Technol*, 6:5 (1983), pp. 471–8.
9. H. H. Chang, M. Hemberg, M. Barahona, D. E. Ingber and S. Huang, 'Transcriptome-Wide Noise Controls Lineage Choice in Mammalian Progenitor Cells', *Nature*, 453:7194 (2008), pp. 544–7; I. Golding, J. Paulsson, S. M. Zawilski and E. C. Cox. 'Real-Time Kinetics of Gene Activity in Individual Bacteria', *Cell*, 123:6 (2005), pp. 1025–36.
10. Golding, Paulsson et al., 'Real-Time Kinetics of Gene Activity in Individual Bacteria', pp. 1025–36.
11. P. C. Nowell, 'The Clonal Evolution of Tumor Cell Populations', *Science*, 194:4260 (1976), pp. 23–8.
12. H. H. Chang, P. Y. Oh, D. E. Ingber and S. Huang, 'Multistable and Multistep Dynamics in Neutrophil Differentiation', *BMC Cell Biol*, 7:1 (2006), p. 11.
13. J. M. Raser and E. K. O'Shea, 'Noise in Gene Expression: Origins, Consequences, and Control', *Science*, 309:5743 (2005), pp. 2010–3.
14. I. Prigogine, *The End of Certainty* (New York: Free Press; 1997).
15. J. Monod and F. Jacob, 'Teleonomic Mechanisms in Cellular Metabolism, Growth, and Differentiation', *Cold Spring Harb Symp Quant Biol*, 26 (1961), pp. 389–401.
16. S. Huang, 'Systems Biology of Stem Cells: Three Useful Perspectives to Help Overcome the Paradigm of Linear Pathways', *Philos Trans R Soc Lond B Biol Sci*, 366:1575 (2011), pp. 2247–59.
17. D. Kaplan and L. Glass, *Understanding Nonlinear Dynamics* (New York: Springer,

1995).

18. J. X. Zhou, M. D. Aliyu, E. Aurell and S. Huang, 'Quasi-Potential Landscape in Complex Multi-Stable Systems', *Journal of the Royal Society, Interface/The Royal Society*, 9:77 (2012), pp. 3539–53.

19. T. S. Gardner, C. R. Cantor and J. J. Collins, 'Construction of a Genetic Toggle Switch in Escherichia Coli', *Nature*, 403:6767 (2000), pp. 339–42.

20. M. Delbrück, Discussion. Unités biologiques douées de continuité génétique Colloques Internationaux du Centre National de la Recherche Scientifique (Paris, CNRS: 1949), pp. 33–5.

21. Huang, 'Systems Biology of Stem Cells: Three Useful Perspectives to Help Overcome the Paradigm of Linear Pathways'.

22. S. Kauffman, 'Homeostasis and Differentiation in Random Genetic Control Networks', *Nature*, 224:215 (1969), pp. 177–8.

23. P. Trojer and D. Reinberg, 'Histone Lysine Demethylases and their Impact on Epigenetics', *Cell*, 125:2 (2006), pp. 213–7.

24. B. Goodwin, *How the Leopard Changed Its Spots: The Evolution of Complexity* (1993; Princeton, NJ: Princeton University Press; 2001).

25. R. L. Coffman and S. L. Reiner, 'Instruction, Selection, or Tampering with the Odds?', *Science*, 284:5418 (1999), pp. 1283–5; T. Enver, C. M. Heyworth and T. M. Dexter, 'Do Stem Cells Play Dice?', *Blood*, 92:2 (1998), pp. 348–51; discussion 52.

26. L. J. Fairbairn, G. J. Cowling, B. M. Reipert and T. M. Dexter, 'Suppression of Apoptosis Allows Differentiation and Development of a Multipotent Hemopoietic Cell Line in the Absence of Added Growth Factors', *Cell*, 74:5 (1993), pp. 823–32.

27. Chang, Hemberg et al., 'Transcriptome-Wide Noise Controls Lineage Choice in Mammalian Progenitor Cells', pp. 544–7.

28. T. Enver, C. M. Heyworth and T. M. Dexter, 'Do Stem Cells Play Dice?', *Blood*, 92:2 (1998), pp. 348–51; discussion 52.

29. Coffman and Reiner, 'Instruction, Selection, or Tampering with the Odds?', pp. 1283–5.

4 Accoto, Guglielmelli and Laschi, Embodied Intelligence in the Biomechatronic Design of Robots

1.   A – Biomedical Robotics and Biomicrosystems Lab, Università Campus Bio-Medico di Roma, Via A. Del Portillo, 21 – 00128 Roma, Italy. B – The BioRobotics Institute, Scuola Superiore Sant'Anna, viale Rinaldo Piaggio 34, 56025 Pontedera (Pisa), Italy

2.   J. F. Engelberger, *Robotics in Service* (Cambridge, MA: Massachusetts Institute of Technology Press, 1989).

3.   For instance, if we admit that biological organisms are optimized for their ecological niche, then we can also conclude that they are the fittest to withstand small changes in the properties of that niche.

4.   R. A. Brooks, 'New Approaches to Robotics', *Science*, 253 (1991), pp. 1227–32; R. C. Arkin, *Behavior-Based Robotics* (Cambridge, MA: Massachusetts Institute of Technology Press, 1998).

5.   R. Pfeifer, M. Lungarella and F. Iida, 'Self-Organization, Embodiment, and Biologically Inspired Robotics', *Science*, 318:5853 (2007), pp. 1088–93; J. F. V. Vincent, O. A. Bogatyreva, N. R. Bogatyrev, A. Bowyer, and A. K. Pahl, 'Biomimetics: Its Prac-

tice and Theory', *Journal of the Royal Society Interface,* 3 (2006), on p. 471.

6. A. Menciassi and C. Laschi, 'Biorobotics', in O. Ziad Abu-Faraj (ed.), *Handbook of Research on Biomedical Engineering Education and Advanced Bioengineering Learning: Interdisciplinary Concepts* (Hershey, PA: IGI Global, 2012).

7. R. Cordeschi, *The Discovery of the Artificial: Behavior, Mind and Machines before and beyond Cybernetics* (Dordrecht: Kluwer, 2002).

8. J. Loeb, *Forced Movements, Tropisms, and Animal Conduct* (Philadelphia, PA, and London: Lippincot, 1918); H. S. Jennings, *Behavior of the Lower Organisms* (New York: Columbia University Press, 1906).

9. E. Datteri and G. Tamburrini, 'Bio-Robotic Experiments and Scientific Method', in L. Magnani and R. Dossena (eds), *Computing, Philosophy and Cognition* (London: College Publications, 2005).

10. B. Webb, *Biorobotics* (Cambridge, MA: Massachusetts Institute of Technology Press, 2001); P. Dario, M. C. Carrozza, E. Guglielmelli, C. Laschi, A. Menciassi, S. Micera and F. Vecchi, 'Robotics as a Future and Emerging Technology: Biomimetics, Cyberneticsand Neuro-Roboticsin European Projects', *IEEE Robotics and Automation Magazine,* 12:2 (2005), pp. 29–43.

11. N. Di Stefano and G. Ghilardi, 'Embodied Intelligence: Epistemological Remarks on an Emerging Paradigm in the Artificial IntelligenceDebate', *Epistemologia,* 36:1 (2013), pp. 100–11.

12. C. Paul, 'Morphological Computation: A Basis for the Analysis of Morphology and Control Requirements', *Robotics and Autonomous Systems,* 54:8 (2006), pp. 619–30.

13. R. Pfeifer and F. Lida, 'Morphological Computation: Connecting Body, Brain and Environment', *Japanese Scientific Monthly,* 58:2 (2005), pp. 48–54.

14. G. Sumbre, Y. Gutfreund, G. Fiorito, T. Flash and B. Hochner, 'Control of Octopus Arm Extension by a Peripheral Motor Program', *Science,* 293 (2001), pp. 1845–8.

15. G. Sumbre, G. Fiorito, T. Flash and B. Hochner, 'Octopuses Use a Human-Like Strategy to Control Precise Point-to-Point Arm Movements', *Curr Biol.,* 16:8 (2006), pp. 767–72.

16. M. Calisti, M. Giorelli, G. Levy, B. Mazzolai, B. Hochner, C. Laschi and P. Dario, 'An Octopus-Bioinspired Solution to Movement and Manipulation for Soft Robots', *Bioinspiration & Biomimetics,* 6:3 (2011), pp. 1–10.

17. L. Margheri, C. Laschi and B. Mazzolai, 'Soft Robotic Arm Inspired by the Octopus: I. From Biological Functions to Artificial Requirements', *Bioinspir. Biomim,* 7:2 (2012), 025004, pp. 1–12.

18. Calisti et al., 'An Octopus-Bioinspired Solution to Movement and Manipulation for Soft Robots'.

19. F. Giorgio-Serchi, A. Arienti and C. Laschi, 'Biomimetic Vortex Propulsion: Towards the New Paradigm of Soft Unmanned Underwater Vehicles', *IEEE/ASME Transactions on Mechatronics,* 18:3 (2013), pp. 1–10.

20. S. Kim, C. Laschi, and B. Trimmer, 'Soft Robotics: A Bioinspired Evolution in Robotics', *Trends in Biotechnology,* 31 (2013), pp. 287–94.

21. L. Pons, *Wearable Robots: Biomechatronic Exoskeletons,* 1st edn (Chichester: John Wiley & Sons, 2008).

22. J. C. R. Licklider, 'Man–Computer Symbiosis', *IRE Transactions on Human Factors in Electronics,* 1 (1960), pp. 4–11.

23. K. Kawamura, T. E. Rogers, K. A. Hambuchen and D. Erol, 'Towards a Human–Robot Symbiotic System', *Robotics and Computer Integrated Manufacturing,* 19 (2003), pp.

555–65.

24. D. M. Wilkes, A. Alford, M. E. Cambron, T. E. Rogers, R. A. Peters and K. Kawamura , 'Designing for Human–Robot Symbiosis', *Industrial Robot*, 26 (1999), pp. 49–58.

25. K. Kawamura, S. Bagchi, M. Iskarous and M. Bishay, 'Intelligent Robotic Systems in Service of the Disabled', *IEEE Transactions on Rehabilitation Engineering*, 3 (1995), pp. 1–9.

26. C. M. Hogan and E. Monosson, 'Symbiosis', *Encyclopedia of Earth*, 106 (2010).

27. F. Sergi, D. Accoto, N. L. Tagliamonte, G. Carpino and E. Guglielmelli, 'A Systematic Graph-Based Method for the Kinematic Synthesis of Non-Anthropomorphic Wearable Robots for the Lower Limbs', *Frontiers of Mechanical Engineering*, 6:1 (2011), pp. 61–70.

28. N. L. Tagliamonte and D. Accoto, 'Passivity Constraints for the Impedance Control of Series Elastic Actuators', *Proceedings of the Institution of Mechanical Engineers, Part I: Journal of Systems and Control Engineering* (2013), pp. 138–53.

29. K. Zuse, 'Computing Space' (1967)

5 MacLeod, Managing Complexity: Model-Building in Systems Biology and Its Challenges for Philosophy of Science

1. See M. A. O'Malley and J. Dupré, 'Fundamental Issues in Systems Biology', *BioEssays*, 27:12 (2005), pp.1270–6.

2. This research is funded by the US National Science Foundation grant DRL097394084. Human subjects restrictions require that we do not identity specific researchers, who are mainly PhD students, and so we refer to the labs with letter designations and lab members with letter plus number.

3. Many of the researchers are non-native English speakers, which accounts for the grammatical errors in the quotes.

4. M. A. O'Malley and J. Dupré, 'Fundamental Issues in Systems Biology', *BioEssays*, 27:12 (2005), pp.1270–6.

5. See H. Kitano, 'Looking Beyond the Details: A Rise in System-Oriented Approaches in Genetics and Molecular Biology', *Current Genetics*, 41:1 (2002), pp.1–10; U. Krohs and W. Callebaut, 'Data without Models Merging with Models without Data', in F. Boogerd, F. J. Bruggeman, J.-S. S. Hofmeyer and H. V. Westerhoff (eds), *Systems Biology: Philosophical Foundations* (Amsterdam: Elsevier, 2007), pp. 181–213; O'Malley and Dupré, 'Fundamental Issues in Systems Biology'.

6. Kitano, 'Looking Beyond the Details: A Rise in System-Oriented Approaches in Genetics and Molecular Biology'; H.V. Westerhoff and D.B. Kell, 'The Methodologies of Systems Biology', in F. Boogerd, F. J. Bruggeman, J.-S.S. Hofmeyer and H. V. Westerhoff (eds), *Systems Biology: Philosophical Foundations* (Amsterdam: Elsevier, 2007), pp. 23–70.

7. L. Hood, J. R. Heath, M. E. Phelps and B. Lin, 'Systems Biology and New Technologies Enable Predictive and Preventative Medicine', *Science Signaling*, 306:5696 (2004), pp. 640–3, on p. 640.

8. U. Alon, *An Introduction to Systems Biology: Design Principles of Biological Circuits* (London: CRC Press, 2006); U. Alon, 'Network Motifs: Theory and Experimental Approaches', *Nature Reviews Genetics*, 8:6 (2007), pp. 450–61.

9. H. V. Westerhoff and D. B. Kell, 'The Methodologies of Systems Biology'; U. Krohs

and W. Callebaut, 'Data without Models Merging with Models without Data', in F. Bo-ogerd, F. J. Bruggeman, J.-S. S. Hofmeyer and H. V. Westerhoff (eds), *Systems Biology: Philosophical Foundations* (Amsterdam: Elsevier, 2007), pp. 181–213; F. J. Bruggeman and H. V. Westerhoff, 'The Nature of Systems Biology', *TRENDS in Microbiology,* 15:1 (2007), pp. 45–50.

10. There are emerging concepts of mesoscopic modelling and mesoscopic levels of organi-zation being developed. See for instance M. Bertolaso, 'Hierarchies and Causal Rela-tionships in Interpretative Models of the Neoplastic Process', *History & Philosophy of the Life Sciences,* 33:4 (2011), pp.515–35; M. Bertolaso, A. Giuliani and S. Filippi, 'The Mesoscopic Level and its Epistemological Relevance in Systems Biology', in A. X. C. N. Valente, A. Sarkar and Y. Gao (eds), *Recent Advances in Systems Biological Research* (Hauppauge, NY: Nova Science Publishers, 2014), pp. 19–36; A. Giuliani, S. Filippi and M. Bertolaso, 'Why Network Approach Can Promote a New Way of Thinking in Biology', *Frontiers in Genetics,* 5:83 (2014), doi:10.3389/fgene.2014.00083. These authors develop what I would consider a more ontological or causal concept which argues for the mesoscopic level as the level or scale of network organization at which functionality emerges in responses to higher-level system and environmental con-straints. Myself and Nersessian develop the concept provided by Voit. See E. O. Voit, 'Mesoscopic Modelling as a Starting Point for Computational Analyses of Cystic Fibro-sis as a Systemic Disease', *Biochimica et Biophysica Acta (BBA)-Proteins and Proteomics,* 1844:1(January 2014), pp. 258–70; E. O. Voit, Q. Zhen and S. Kikuchi, 'Mesoscopic Models of Neurotransmission as Intermediates between Disease Simulators and Tools for Discovering Design Principles', *Pharmacopsychiatry,* 45:1 (2012), pp. 22–30, on p.23. This concept has a distinct epistemic or investigative dimension, being concerned with how to transition to fuller scale network representations or to design principle analysis from cognitively manageable network sizes. The concepts, however, undoubt-edly overlap, since mesoscopic models on Voit's account, and those of our modellers, pick out subsystems that account for adequately at least particular functions and other phenomena they are interested in. These subsystems can thus be construed as essential units of functionality at the middle-level.

11. D. Noble, *The Music of Life: Biology Beyond Genes* (New York: Oxford University Press, 2008), p. 79.

12. Voit, 'Mesoscopic Modelling as a Starting Point for Computational Analyses of Cystic Fibrosis as a Systemic Disease'.

13. Voit, Zhen and Kikuchi, 'Mesoscopic Models of Neurotransmission as Intermediates between Disease Simulators and Tools for Discovering Design Principles', p. 23.

14. W. Bechtel, 'Looking Down, Around, and Up: Mechanistic Explanation in Psychol-ogy', *Philosophical Psychology,* 22:5 (2009), pp. 543–64.

15. M.A. Savageau, 'Biochemical Systems Analysis: I. Some Mathematical Properties of the Rate Law for the Component Enzymatic Reactions', *Journal of theoretical Biology,* 25:3 (1969), pp. 365–9; E. O. Voit, *Computational Analysis of Biochemical Systems: A Practi-cal Guide for Biochemists and Molecular Biologists* (Cambridge: Cambridge University Press: 2000).

16. Bechtel, 'Looking Down, Around, and Up'.

17. Voit, *Computational Analysis of Biochemical Systems*

18. C. F. Craver, *Explaining the Brain: What a Science of the Mind-Brain Could Be* (New York: Oxford University Press, 2007); W. Bechtel, 'The Downs and Ups of Mechanistic Research: Circadian Rhythm Research as an Exemplar', *Erkenntnis,* 73:3 (2010), pp.

313–28; W. Bechtel and R. Richardson, *Discovering Complexity: Decomposition and Localization as Strategies in Scientific Research* (Cambridge, MA: Massachusetts Institute of Technology Press, 2010).

19. O'Malley and Dupré, 'Fundamental Issues in Systems Biology', *BioEssays*, 27:12 (2005), pp. 1270–6.
20. O'Malley and Dupré, 'Fundamental Issues in Systems Biology', p. 1273.
21. See, for instance, H. V. Westerhoff and D. B. Kell, 'The Methodologies of Systems Biology' in F. Boogerd, F. J. Bruggemanm, J.-S. S. Hofmeyer and H. V. Westerhoff (eds), *Systems Biology: Philosophical Foundations* (Amsterdam: Elsevier, 2007), pp. 23–70.
22. See M. MacLeod and N.J. Nersessian, 'Strategies for Coordinating Experimentation and Modelling in Integrative Systems Biology', *J. Exp. Zool. (Mol. Dev. Evol.)*, 9999 (2014), pp. 1–10.
23. This is a particularly interesting issue we've also studied. At the present the situation seems quite mixed. Some molecular biologists, particularly those with modelling experience or education, even in partial amounts, seem well disposed to working with modellers and have appropriate expectations when entering collaborative relationships. Many molecular biologists however are ambivalent towards systems biology or quite dismissive of it. Some of the collaborative relationships fail to succeed because these kinds of attitudes were playing a role. There is strong tendency in the molecular biology community to portray systems biology as just sophisticated data-fitting but largely unproductive for their needs. If this attitude can't be shifted there will be significant obstacles to systems biology getting the kinds of experimental resources it needs.
24. M. MacLeod and N.J. Nersessian, 'Coupling Simulation and Experiment: The Bimodal Strategy in Integrative Systems Biology', *Studies in History and Philosophy of Biological and Biomedical Sciences*, 44 (2013), pp. 572–84.
25. S.T. Kuhn, *The Structure of Scientific Revolutions* (Chicago, IL: University of Chicago Press, 1962).
26. M. MacLeod and N.J. Nersessian, 'Building Simulations from the Ground-Up: Modelling and Theory in Systems Biology', *Philosophy of Science*, 80:4 (2013), pp. 533–56.
27. MacLeod and Nersessian, 'Building Simulations from the Ground-Up: Modelling and Theory in Systems Biology'.
28. M. Weisberg and R. Muldoon, 'Epistemic Landscapes and the Division of Cognitive Labor', *Philosophy of Science*, 76:2 (2009), pp. 225–52.
29. P. Machamer, L. Darden and C. F. Craver, 'Thinking about Mechanisms', *Philosophy of Science*, 67:1 (2000), pp. 1–25.
30. W. Bechtel and A. Abrahamsen, 'Explanation: A mechanist alternative', *Studies in History and Philosophy of Science Part C: Studies in History and Philosophy of Biological and Biomedical Sciences*, 36:2 (2005), pp. 421–41; W. Bechtel and A. Abrahamsen, 'Dynamic Mechanistic Explanation: Computational Modelling of Circadian Rhythms as an Exemplar for Cognitive Science', *Studies in History and Philosophy of Science Part A*, 41:3 (2010), pp. 321–33.
31. I. Brigandt, 'Systems Biology and the Integration of Mechanistic Explanation and Mathematical Explanation', *Studies in History and Philosophy of Biological and Biomedical Sciences*, 44:4 (2013), pp. 477–92.
32. H. Kitano, 'Looking Beyond the Details: A Rise in System-Oriented Approaches in Genetics and Molecular Biology', *Current Genetics*, 41:1 (2002), pp. 1–10, on p .1.
33. Brigandt, 'Systems Biology and the Integration of Mechanistic Explanation and Mathematical Explanation'; A. Levy and W. Bechtel, 'Abstraction and the Organization of

Mechanisms', *Philosophy of Science*, 80:2 (2013), pp. 241–61.

34. M. MacLeod and N.J. Nersessian, unpublished; N.J. Nersessian and W. C. Newstetter, 'Interdisciplinarity in Engineering Research and Learning', in A. Jori and B. Olds (eds), *Cambridge Handbook of Engineering Education Research* (Cambridge: Cambridge University Press, 2013), pp. 713–30.

35. M. MacLeod and N. J. Nersessian, 'Coupling Simulation and Experiment: The Bimodal Strategy in Integrative Systems Biology', *Studies in History and Philosophy of Biological and Biomedical Sciences*, 44 (2013), pp. 572–84.

36. W. C. Wimsatt, *Re-Engineering Philosophy for Limited Beings: Piecewise Approximations to Reality* (Cambridge, MA: Harvard University Press, 2007).

37. H. M. Collins and R. Evans, 'The Third Wave of Science Studies Studies of Expertise and Experience', *Social Studies of Science*, 32:2 (2002), pp. 235–96.

6 Boem, Boniolo and Pavelka, Stratification and Biomedicine: How Philosophy Stems from Medicine and Biotechnology

1. We wish to thank the students and the staff of the PhD programme in 'Foundations of the Life Sciences and their Ethical Consequences' at the European School of Molecular Medicine (SEMM, Milan) and all the members of the Biomedical Humanities Unit of the Department of Experimental Oncology (Istituto Europeo di Oncologia, IEO, Milan). Z. Pavelka has worked especially on the section entitled 'Stratification Medicine'. F. Boem has worked mainly on the section entitled 'The Philosophy Within'. G. Boniolo has supervised the entire work.

2. D. C. Whitcomb, 'What Is Personalized Medicine and What Should It Replace?', *Nature Reviews. Gastroenterology & Hepatology*, 9:7 (2012), pp. 418–24, on pp. 418, 421–2; National Research Council, 'Toward Precision Medicine: Building a Knowledge Network for Biomedical Research and a New Taxonomy of Disease'. (2011), at http://www.nap.edu/openbook.php?record_id=13284. [accessed 5 May 2014]; European Science Foundation., 'Personalised Medicine for the European Citizen : European Science Foundation', 2012, at http://www.esf.org/coordinating-research/forward-looks/biomedical-sciences-med/current-forward-looks-in-biomedical-sciences/personalised-medicine-for-the-european-citizen.html [accessed 30 April 2014].

3. D. C. Whitcomb, 'What Is Personalized Medicine and What Should It Replace?', pp. 418, 421–2; National Research Council, 'Toward Precision Medicine: Building a Knowledge Network for Biomedical Research and a New Taxonomy of Disease'; European Science Foundation., 'Personalised Medicine for the European Citizen : European Science Foundation'.

4. A. J. Atkinson, A.C. Wayne, V. G. DeGruttola, D. L. DeMets, G. J. Downing, D. F. Hoth, J. A. Oates et al., 'Biomarkers and Surrogate Endpoints: Preferred Definitions and Conceptual Framework', *Clinical Pharmacology & Therapeutics*, 69:3 (2001), pp. 89–95, on p. 91.

5. J. J. Lee, 'Chapter 20 – Statistical Methods for Biomarker Analysis for Head and Neck Carcinogenesis and Prevention', in F. John, J. Ensley, S. Gutkind, R. J. Jacobs and S. M. Lippman (eds), *Head and Neck Cancer*, 287–IV (San Diego, CA: Academic Press, 2003), p. 288.

6. K. Bracht, 'Biomarker: Indikatoren für Diagnose und Therapie', *Pharmazeutische Zeitung*, 12/2009 (March 2009), at http://www.pharmazeutische-zeitung.de/index.

php?id=29346. [accessed 3 April 2014].

7.   I. Koychev, E. Barkus, U. Ettinger, S. Killcross, J. P. Roiser, L. Wilkinson and B. Deakin, 'Evaluation of State and Trait Biomarkers in Healthy Volunteers for the Development of Novel Drug Treatments in Schizophrenia', *Journal of Psychopharmacology*, 25:9 (2011), pp. 1207–25, on p. 1207; R. S. Vasan, 'Biomarkers of Cardiovascular Disease Molecular Basis and Practical Considerations', *Circulation*, 113:19 (2006), pp. 2335–62, on p. 2335.

8.   Bracht, 'Biomarker: Indikatoren für Diagnose und Therapie'.

9.   See WHO, 'Priority Medicines for Europe and the World Update Report, 2013', 2013, at http://www.who.int/medicines/areas/priority_medicines/en/ [accessed 30 April 2014].

10. W. E. Evans and M. V. Relling., 'Pharmacogenomics: Translating Functional Genomics into Rational Therapeutics', *Science*, 286:5439 (1999), pp. 487–91, on pp. 487–8.

11. D. C. Whitcomb, 'What Is Personalized Medicine and What Should It Replace?', *Nature Reviews. Gastroenterology & Hepatology*, 9:7 (2012), pp. 418–24, on pp. 418, 421–2; R. L. Schilsky, 'Personalized Medicine in Oncology: The Future Is Now', *Nature Reviews. Drug Discovery*, 9:5 (2010), pp. 363–6, on p. 365.

12. H. C. A. Kraemer, E. Kazdin, D. R. Offord, R. C. Kessler, P. S. Jensen and D. J. Kupfer., 'Coming to Terms with the Terms of Risk', *Archives of General Psychiatry*, 54:4 (1997), pp. 337–43, on p. 338.

13. J. G. Dreyfus, P. L. Lutsey, R. Huxley, J. S. Pankow, E. Selvin, L. Fernández-Rhodes, N. Franceschini and E. W. Demerath., 'Age at Menarche and Risk of Type 2 Diabetes among African-American and White Women in the Atherosclerosis Risk in Communities (ARIC) Study', *Diabetologia*, 55:9 (2012), pp. 2371–80.

14. Psychiatric GWAS Consortium Coordinating Committee, S. Cichon, N. Craddock, M. Daly, S. V. Faraone, P. V. Gejman, J. Kelsoe et al., 'Genomewide Association Studies: History, Rationale, and Prospects for Psychiatric Disorders', *American Journal of Psychiatry*, 166:5 (2009), pp. 540–56.

15. Psychiatric GWAS Consortium Coordinating Committee et al., 'Genomewide Association Studies: History, Rationale, and Prospects for Psychiatric Disorders'.

16. K. Bettens, K. Sleegers and C. Van Broeckhoven, 'Current Status on Alzheimer Disease Molecular Genetics: From Past, to Present, to Future', *Human Molecular Genetics*, 19:1 (2010), pp. R4–R11, on p. R6.

17. R. Chen and M. Snyder, 'Promise of Personalized Omics to Precision Medicine', *Wiley Interdisciplinary Reviews: Systems Biology and Medicine*, 5:1 (2013), pp. 73–82.

18. L. M. R. Ferreira and M. A. Mostajo-Radji, 'How Induced Pluripotent Stem Cells Are Redefining Personalized Medicine', *Gene*, 520:1 (2013), pp. 1–6, on p. 4.

19. See B. J. Strasser, 'Genetics. GenBank – Natural History in the 21st Century?', *Science (New York, N.Y.)*, 322:5901 (2008), pp. 537–8 and B. J. Strasser, 'Data-Driven Sciences: From Wonder Cabinets to Electronic Databases', *Studies in History and Philosophy of Biological and Biomedical Sciences*, 43:1 (2012), pp. 85–7.

20. See G. Boniolo, *On Scientific Representation: From Kant to a New Philosophy of Science*, 1st edn (Basingstoke and New York: Palgrave Macmillan, 2007).

21. H. Poincaré, *Science and Hypothesis*, trans. W. J. Greenstreet (1901; New York: Cosimo Classics, 2007).

22. See S. Brenner, 'Sequences and Consequences', *Philosophical Transactions of the Royal Society B: Biological Sciences*, 365:1537 (2010), pp. 207–12.

23. M. A. O'Malley and O. S. Soyer, 'The Roles of Integration in Molecular Systems

Biology', *Studies in History and Philosophy of Science Part C: Studies in History and Philosophy of Biological and Biomedical Sciences*, 43:1 (2012), pp. 58–68, on p.60.

24. M. A. O'Malley and O. S. Soyer, 'The Roles of Integration in Molecular Systems Biology', *Studies in History and Philosophy of Science Part C: Studies in History and Philosophy of Biological and Biomedical Sciences*, 43:1 (2012), pp. 58–68, on p. 60.

25. O'Malley and Soyer, 'The Roles of Integration in Molecular Systems Biology'.

26. T. R. Gruber, 'A Translation Approach to Portable Ontology Specifications', *Knowl. Acquis*, 5:2 (1993), pp. 199–220.

27. *Encyclopedia of Database Systems.*, 2014, at http://www.springer.com/computer/data base+management+%26+information+retrieval/book/978-0-387–35544-3. (2014) [accessed 5 May 2014].

28. S. Leonelli, 'Bio-Ontologies as Tools for Integration in Biology', *Biological Theory*, 3:1 (2008), pp. 7–11, on p. 1.

29. B. Smith, M. Ashburner, C. Rosse, J. Bard, W. Bug, W. Ceusters, L. J. Goldberg et al., 'The OBO Foundry: Coordinated Evolution of Ontologies to Support Biomedical Data Integration', *Nature Biotechnology*, 25:11 (2007), pp. 1251–55, on pp. 1251–3.

30. B. Smith et al., 'The OBO Foundry: Coordinated Evolution of Ontologies to Support Biomedical Data Integration'.

31. B. Smith, 'Barry Smith: Introduction to Biomedical Ontology – Streaming Video', 2008, at http://ontology.buffalo.edu/smith/BioOntology_Course.html. [accessed 5 May 2014].

32. B. C. Van Fraassen, *Scientific Representation: Paradoxes of Perspective* (Oxford: Oxford University Press, 2008), p. 2.

33. See the classical E. A. Burtt, *The Metaphysical Foundations of Modern Physical Science: A Historical and Critical Essay* (London: Keagan Paul, Trench, Trubner, 1923), at http://archive.org/details/metaphysicalfoun00burtuoft. [accessed 20 April 2014].

34. K. Popper, *The Logic of Scientific Discovery*, 1st edn (London: Routledge, 1963).

35. G. Boniolo, *On Scientific Representation: From Kant to a New Philosophy of Science*, 1st edn (Basingstoke and New York: Palgrave Macmillan, 2007), ch. 1.

7 Ghilardi, Epistemology of Robotics: An Outline

1. K. Čapek, *R.U.R.*, trans. P. Selver and N. Playfair (1920; Mineola, NY: Dover Pub. Inc., 2001).

2. http://www.britannica.com/EBchecked/topic/194023/ethics [accessed 20 July 2014].

3. W. Wallach and C. Allen, *Moral Machines, Teaching Robots Right from Wrong* (Oxford: Oxford University Press, 2008).

4. Y. Sherwin, 'Machine Morality: Computing Right and Wrong', *Yale Scientific*, 10 May 2012, at http://www.yalescientific.org/2012/05/machine-morality-computing-right-and-wrong/ [accessed 20 July 2014].

5. Sherwin, 'Machine Morality: Computing Right and Wrong'.

6. C. T. Rubin, 'Machine Morality and Human Responsibility', *New Atlantis* (22 April 2011), pp. 58–79.

7. N. Wiener, *God and Golem: A Comment on Certain Points where Cybernetics Impinges on Religion* (Cambridge, MA: Massachusetts Institute of Technology Press, 1966), p. 11.

8. On this topic see R. Pfeifer and J. Bongard, *How the Body Shapes the Way We Think: A New View of Intelligence* (London: Massachusetts Institute of Technology Press, 2007); N. Di Stefano and G. Ghilardi 'Embodied Intelligence: Epistemological Remarks on an

Emerging Paradigm in the Artificial Intelligence Debate', *Epistemologia*, 36, 2013, pp. 100–11.

9.　A. Turing, 'Computing Machinery and Intelligence', *Mind*, 59 (1950), pp. 433–60.

10.　Functionalism is the doctrine that what makes something a thought, desire, pain (or any other type of mental state) depends not on its internal constitution, but solely on its function, or the role it plays, in the cognitive system of which it is a part.

11.　On this theme consider P. Lin, J. Beckey and K. Abney, 'Autonomous Military Robotics: Risk, Ethics, and Design', *Ethics, Emerging Sciences Group* (San Luis Obispo, CA: Polytechnic State University, 2008), p. 65.

12.　P. Lin, J. Beckey and K. Abney, 'Autonomous Military Robotics: Risk, Ethics, and Design', *Ethics, Emerging Sciences Group* (San Luis Obispo, CA: Polytechnic State University, 2008), p. 65.

13.　The modern father of this concept is L. Kohlberg, *Essays on Moral Development: Vol. 2. The Psychology of Moral Development* (San Francisco, CA: Harper & Row, 1984). For a critical exposition of instrumental morality see D. L. Krebs and K. Denton, 'Toward a More Pragmatic Approach to Morality: A Critical Evaluation of Kohlberg's Model', *American Psychological Association*, 112:3 (2005), pp. 629–49.

14.　G. N. Sandor and A. G. Erdman, *Advanced Mechanism Design*, Vol. I – II (Englewood Cliffs, NJ: Prentice-Hall, 1984), p. 16.

15.　Aristotle, *Nicomachean Ethics*, 6, 11, 1143 a35–b3.

16.　A robot may not injure a human being or, through inaction, allow a human being to come to harm. A robot must obey the orders given to it by human beings, except where such orders would conflict with the First Law. A robot must protect its own existence as long as such protection does not conflict with the First or Second Laws. These laws appear for the first time in the short story, 'Runaround', in I. Asimov, *I, Robot* (New York: Doubleday & Company, 1950).

17.　Y. Sherwin, 'Machine Morality: Computing Right and Wrong', *Yale Scientific*, (2012), at http://www.yalescientific.org/2012/05/machine-morality-computing-right-and-wrong [acccessed 20 July 2014].

18.　Isaac Asimov from the introduction to *The Rest of the Robots* (Frogmore, Sant Albans: Granada Publishing Limited, Panther Books Ltd., 1968), p. 13.

19.　I. Kant, *The Science of Right* (1790), in *The Philosophy of Law, an Exposition of the Fundamental Principles of Jurisprudence, as the Science of Right by Immanuel Kant*.

20.　Kant, *The Science of Right*, pp. 46–8.

21.　R. Pfeifer and J. Bongard, *How the Body Shapes the Way We Think: A New View of Intelligence* (London: MIT Press, 2007).

22.　R. Pfeifer, 'Building Fungus Eaters: Design Principles of Autonomous Agents', in *Proceedings of the third Simulation of Adaptive Behavior Conference* (Cambridge, MA: Massachusetts Institute of Technology Press, 1996).

23.　Pfeifer, 'Building Fungus Eaters: Design Principles of Autonomous Agents'.

24.　Pfeifer, 'Building Fungus Eaters: Design Principles of Autonomous Agents'.

25.　Pfeifer and Bongard, *How the Body Shapes the Way We Think*, p. 140.

26.　Pfeifer and Bongard, *How the Body Shapes the Way We Think*, p. 176.

27.　M. Czikszentmihalyi, *Flow: The Psychology of Optimal Experience* (New York: Harper and Row, 1990).

28.　Pfeifer and Bongard, *How the Body Shapes the Way We Think*, p. 168.

29.　Pfeifer and Bongard, *How the Body Shapes the Way We Think*, p. 137

30.　On this topic see S. Gallagher, 'Moral Agency, Self-Consciousness, and Practical Wisdom', *Journal of Consciousness Studies*, 14:5–6 (2007), pp. 199–223.

31. Gallagher, 'Moral Agency, Self-Consciousness, and Practical Wisdom', p. 206
32. We assume the difference between computing and calculating, the former is kind of comparison, while the latter implies an enumeration.
33. On this subject see F. Varela and J. Shear, *The View from Within* (Thorverton: Imprint Academic, 1999).
34. See E. Thamiry, 'Immanence' (1910), *The Catholic Encyclopedia, Volume VII* (Robert Appleton Company, online edition, copyright Kevin Knight, 1999).
35. On this subject see J. Koszteyn, 'Actio Immanens – A Fundamental Concept of Biological Investigation', *Forum Philosophicum*, Krakow, 8 (2003), pp. 81–120; J. J. Sanguineti, 'Immanenza e transitività nell'operare umano', Proceedings of the III International Congress SITA, *Etica e società contemporanea*, Vol. I (Rome: Ed. Vaticana, 1992), pp. 261–74.
36. Koszteyn, 'Actio Immanens'; Sanguineti, 'Immanenza e transitività nell'operare umano', p. 81.
37. On this subject see J. Parthemore and B. Whitby, 'What Makes Any Agent a Moral Agent?: Reflections on Machine Consciousness and Moral Agency', *Journal of Machine Consciousness,* (in press), also available at http://projekt.sol.lu.se/fileadmin/user_upload/project/ccs/moralagency.pdf [accessed 20 July 2014].
38. On bio-mechatronics see E. Guglielmelli, L. Zollo and D. Accoto, 'Criteri di progettazione di sistemi robotici per la neuroriabilitazione', in P. Dario, S. Martinoia, G. Rizzolatti and G. Sandini (eds), *Neuro-Robotica, Neuroscienze e robotica per lo sviluppo di macchine intelligenti* (Bologna: Pàtron Ed., 2006).
39. R. A. Brooks, 'Intelligence without Representation', *Artificial Intelligence*, 47 (1991), pp. 139–59.
40. S. Thrun, W. Burgard and D. Fox, *Probabilistic Robotics* (Cambridge, MA: MIT Press, 2005).
41. We agree with Tamburrini where he points out the need to distinguish between the different levels involved in robo-ethics in order to achieve a better understanding of the issues at stake. See G. Tamburrini, 'Robot Ethics: A View from the Philosophy of Science', in R. Capurro and M. Nagenborg (eds), *Ethics and Robotics* (Heidelberg: Akad. Verlagsgesellschaft, 2009).

8 Gonzalez, Prediction and Prescription in Biological Systems: The Role of Technology for Measurement and Transformation

1. The final version of this chapter was prepared at the London School of Economics when the author was a visiting researcher of the Centre for Philosophy of Natural and Social Sciences.
2. This connection between prediction and prescription, which is characteristic of applied science, is different from the relation between explanation and prediction, which belongs to the field of basic science. In this regard, 'historical explanations in natural science cannot be interpreted as potential predictions ... It is clear that scientists sometimes acquire a good understanding of how an event was caused without being able to predict its occurrence. A good example is the inability of seismologists to successfully predict earthquakes even though they are able to explain those that do occur in remarkable detail', C. E. Cleland, 'Prediction and Explanation in Historical Natural Science', *British Journal for the Philosophy of Science*, 62:3 (2011), pp. 566–7.

3. See, in this regard, W. J. Gonzalez, 'Evolutionism from a Contemporary Viewpoint: The Philosophical-Methodological Approach', in W. J. Gonzalez (ed.), *Evolutionism: Present Approaches* (A Coruña: Netbiblo, 2008), pp. 3–59.

4. D. W. McShea, 'Complexity and Evolution: What Everybody Knows', *Biology and Philosophy*, 6 (1991), pp. 303–24, on p. 303. Reprinted in D. Hull and M. Ruse (eds), *The Philosophy of Biology* (Oxford: Oxford University Press, 1998), p. 625.

5. As a matter of fact, some parasites and microorganisms have evolved toward more simplicity. See, in this regard, A. Diéguez, *La vida bajo escrutinio* (Barcelona: Biblioteca Buridán, 2012), pp. 183–5.

6. See C. Darwin, *The Origin of Species by Means of Natural Selection: or the Preservation of Favoured Races in the Struggle for Life* (London: John Murray, 1859). Reprinted by University of Chicago Press, Chicago, IL, 1952, and later on in facsimile with an introduction by Ernst Mayr (Cambridge, MA: Harvard University Press, 1964).

7. See N. Eldredge and S. J. Gould, 'Punctuated Equilibria: An Alternative to Phyletic Gradualism', in T. J. M. Schopf (ed.), *Models in Paleobiology* (San Francisco, CA: Freeman, 1972), pp. 82–115.

8. The concept of 'species' is discussed in A. Marcos, 'The Species Concept in Evolutionary Biology: Current Polemics', in W. J. Gonzalez (ed.), *Evolutionism: Present Approaches* (A Coruña: Netbiblo, 2008), pp. 121–42.

9. Commonly, evolution in biology involves interaction with the environment. On the one hand, there are new species as time passes by, usually because of such interaction; and, on the other, the environment also has clear changes, some due to the intervention of organisms.

10. See, in this regard, McShea, 'Complexity and Evolution: What Everybody Knows', pp. 646–9.

11. See N. Rescher, *Complexity: A Philosophical Overview* (New Brunswick, NJ: Transaction Publishers, 1998), pp. 1–24; especially p. 9.

12. See W. J. Gonzalez, 'The Sciences of Design as Sciences of Complexity: The Dynamic Trait', in H. Andersen, D. Dieks, W. J. Gonzalez, Th. Uebel and G. Wheeler (eds), *New Challenges to Philosophy of Science* (Dordrecht: Springer, 2013), pp. 299–311.

13. See Rescher, *Complexity*, p. 9. His examples have been adapted to the case of biological sciences.

14. See Rescher, *Complexity*, p. 9.

15. 'Morphological complexity of a system (biological or otherwise) is some function of the number of different parts it has', McShea, 'Complexity and Evolution: What Everybody Knows', p. 627.

16. 'Heterogeneous, elaborate, or patternless systems are complex', McShea, 'Complexity and Evolution: What Everybody Knows', p. 627.

17. 'In evolution ... complexity and organization probably *are* connected, because more complex organisms need more organization in order to survive', McShea, 'Complexity and Evolution: What Everybody Knows', p. 627.

18. See, in this regard, A. Potochnick and B. McGill, 'The Limitations of Hierarchical Organizations', *Philosophy of Science*, 79:1 (2012), pp. 120–40.

19. Potochnick and McGill, 'The Limitations of Hierarchical Organizations', p. 121.

20. See W. J. Gonzalez, 'Rationality and Prediction in the Sciences of the Artificial: Economics as a Design Science', in M. C. Galavotti, T. Scazzieri and P. Suppes (eds), *Reasoning, Rationality, and Probability* (Stanford, CA: CSLI Publications, 2008), pp. 165–86; especially pp. 169–71.

21. Rescher, *Complexity*, p. 15.
22. Rescher, *Complexity: A Philosophical Overview*, p. 12. This feature can be related to the analysis of invariant elements in nature.
23. McShea, 'Complexity and Evolution: What Everybody Knows', p. 626.
24. McShea, 'Complexity and Evolution: What Everybody Knows', p. 626.
25. See G. G. Simpson, *The Meaning of Evolution: A Study of the History of Its Significance for Man* (New Haven, CT: Yale University Press, 1949).
26. 'Accurate' is related to the correctness of the prediction made, and 'precision' concerns the level of detail of such a prediction. An analysis of both notions in the context of the social sciences can be seen in W. J. Gonzalez, 'The Role of Experiments in the Social Sciences: The Case of Economics', in T. Kuipers (ed.), *General Philosophy of Science: Focal Issues* (Amsterdam: Elsevier, 2007), pp. 275–301; especially pp. 295–8.
27. The analysis follows here McShea, 'Complexity and Evolution: What Everybody Knows', pp. 628–33. See also S. J. Gould, 'Eternal Metaphors of Paleontology', in A. Hallan (ed.), *Patterns of Evolution* (Amsterdam: Elsevier, 1977), pp. 1–26.
28. The context of his vision is in J. B. Lamarck, [de Monet, Chevalier] de, *Philosophie zoologique, ou, Exposition des considérations relative à l'Histoire naturelle des animaux* (Paris: Chez Dentu [et] L'Auteur, 1809). Translated by Hugh Elliot with introductory essays by D. L. Hull and R. W. Burkhardt: *Zoological Philosophy: An Exposition with Regard to the Natural History of Animals* (Chicago, IL: The University of Chicago Press, 1984).
29. His work has been very influential, especially through J. Maynard Smith, *The Theory of Evolution* (Cambridge: Cambridge University Press, 1993).
30. See G. L. Stebbins, *The Basis of Progressive Evolution* (Chapel Hill, NC: University of North Carolina Press, 1969).
31. See McShea, 'Complexity and Evolution: What Everybody Knows', pp. 633–43.
32. Applied science looks for the resolution of specific problems related to a practical domain. I. Niiniluoto, 'Approximation in Applied Science', *Poznan Studies in the Philosophy of the Sciences and the Humanities*, 42 (1995), pp. 127–39; and I. Niiniluoto, 'The Emergence of Scientific Specialities: Six Models', *Poznan Studies in the Philosophy of the Sciences and the Humanities*, 44 (1995), pp. 211–23.
33. M. Ruse, *Darwinism and Its Discontents* (New York: Cambridge University Press, 2006), p. 209.
34. A philosophico-methodological analysis of Darwinism in the context of evolutionism is in W. J. Gonzalez, 'Evolutionism from a Contemporary Viewpoint: The Philosophical-Methodological Approach', in W. J. Gonzalez (ed.), *Evolutionism: Present Approaches*, pp. 3–59; especially pp. 4–5, 10–21, 30–41.
35. Ruse, *Darwinism and Its Discontents*, p. 209.
36. Ruse, *Darwinism and Its Discontents*, p. 209.
37. On the concept of scientific 'prediction' and the philosophico-methodological views on it see W. J. Gonzalez, *La predicción científica: Concepciones filosófico-metodológicas desde H. Reichenbach a N. Rescher* (Barcelona: Montesinos, 2010); and W. J. Gonzalez, *Philosophico-Methodological Analysis of Prediction and its Role in Economics* (Dordrecht: Springer, forthcoming).
38. These three cases are particularly clear in the case of economics, see W. J. Gonzalez, 'The Evolution of Lakatos's Repercussion on the Methodology of Economics', *HOPOS: Journal of the International Society for the History of Philosophy of Science*, 4:1 (2014), pp. 1–25.

39. See, for example, H. A. Simon, 'Prediction and Prescription in Systems Modeling', *Operations Research*, 38 (1990), pp. 7–14 (reprinted in H. A. Simon, *Models of Bounded Rationality*, Vol. 3: *Empirically Grounded Economic Reason* (Cambridge, MA: MIT Press, 1997), pp. 115–28); H. A. Simon, 'Science Seeks Parsimony, not Simplicity: Searching for Pattern in Phenomena', in A. Zellner, H. A. Keuzenkamp and M. McAleer (eds), *Simplicity, Inference and Modelling: Keeping it Sophisticatedly Simple* (Cambridge: Cambridge University Press, 2001), pp. 32–72; and H. A. Simon, 'Forecasting the Future or Shaping it?', *Industrial and Corporate Change*, 11:3 (2002), pp. 601–5.

40. W. Salmon, 'Rational Prediction', *British Journal for the Philosophy of Science*, 32 (1981), pp. 115–25, on p. 123.

41. A. Wagner, 'The Role of Randomness in Darwinian Evolution', *Philosophy of Science*, 79:1 (2012), pp. 95–119, on p. 108.

42. A. Wagner, 'The Role of Randomness in Darwinian Evolution', p. 108. A prediction can be an inference but not all inference is a prediction. This means that 'prediction' is not a simple 'testable implication'. On this issue, see Gonzalez, *Philosophico-Methodological Analysis of Prediction and its Role in Economics*, chs 2, 8.

43. See, for example, J. Gerhart and M. Kirschner, 'Evolutionary Novelty: The Example of Lactose Synthetase', *Proceedings of the Royal Society of London*, series B, 256 (1994), pp. 53–8. Reprinted in M. Ridley (ed.), *Evolution*, Second edition (Oxford: Oxford University Press, 2004), pp. 326–8.

44. According to Michael R. Dietrich, 'emergent properties are those that arise from the ways that constituent parts interact with one another and are generally considered to be not predictable from simply knowing the constituent parts, and as such constitute phenomena distinct from any lower-level phenomena', M. R. Dietrich, 'Microevolution and Macroevolution are Governed by the Same Processes', in F. J. Ayala and R. Arp (eds), *Contemporary Debates in Philosophy of Biology* (Chichester: Wiley-Blackwell, 2010), p. 173. See also T. Grantham, 'Is Macroevolution More Than Successive Rounds of Microevolution?', *Palaeontology*, 50 (2007), pp. 75–85.

45. See N. Rescher, *Predicting the Future* (New York: State University of New York Press, 1998), pp. 134–5.

46. See W. J. Gonzalez, 'Methodological Universalism in Science and its Limits: Imperialism versus Complexity', in K. Brzechczyn and K. Paprzycka (eds), *Thinking about Provincialism in Thinking*, Poznan Studies in the Philosophy of the Sciences and the Humanities, vol. 100 (Amsterdam and New York: Rodopi, 2012), pp. 155–75.

47. 'In ordinary English, a random event is one without order, predictability or pattern. The word connotes disaggregation, falling apart, formless anarchy, and fear'. This quote from the late Stephen J. Gould (1993) illustrates one reason why many nonbiologists – even highly educated ones – may feel uncomfortable with Darwinian evolution: Darwinian evolution centrally involves chance or randomness', Wagner, 'The Role of Randomness in Darwinian Evolution', p. 95.

48. Rescher, *Complexity: A Philosophical Overview*, p. 8.

49. Besides the philosophico-methodological problem of the limits of the predictability in biological sciences, there is the issue of the present stage of scientific knowledge. This is a historical situation: 'At present, because of difficulties we have canvassed earlier, the ability to predict phenotypes from genotypes is limited', P. Kitcher, *The Lifes to Come: The Genetic Revolution and Human Possibilities* (New York: Simon and Schuster, and London: Penguin, 1996), p. 311. 'Sometimes, predictive tests will inform patients that

they are at high risk for future diseases and, in an unpredictable proportion of these cases, those diagnosed will be able to act to reduce changes', Kitcher, *The Lifes to Come*, p. 182.

50. See, in this regard, W. Parker, 'When Climate Models Agree: The Significance of Robust Model Predictions', *Philosophy of Science*, 78:4 (2011), pp. 579–600.

51. 'Selectionist evolutionary theory has often been faulted for not being able to make novel predictions', R. G. Winther, 'Prediction in Selectionist Evolutionary Theory', *Philosophy of Science*, 76:5 (2009), pp. 889–901, on p. 889.

52. 'The complexity of the situation is, however, so overwhelming that we cannot predict whether or not an environmental challenge will evoke an adaptive evolutionary response in concrete cases. A response will not occur if genetic raw materials for it are unavailable', T. Dobzhansky, 'Chance and Creativity in Evolution', in F. J. Ayala and T. Dobzhansky (eds), *Studies in the Philosophy of Biology, Reduction and Related Problems* (London: Macmillan, 1974), p. 318.

53. There are some similarities with economics. In this regard, see W. J. Gonzalez, 'Prediction and Prescription in Economics: A Philosophical and Methodological Approach', *Theoria*, 13:2 (1998), pp. 321–45.

54. See I. Niiniluoto, 'The Aim and Structure of Applied Research', *Erkenntnis*, 38 (1993), pp. 1–21; especially pp. 9, 19.

55. 'The returns to the application of scientific knowledge to invention might vary across technological fields', L. Fleming and O. Sorenson, 'Science as a Map in Technological Search', *Strategic Management Journal*, 25:8/9 (2004), pp. 909–28, on p. 926.

56. See W. J. Gonzalez, 'The Roles of Scientific Creativity and Technological Innovation in the Context of Complexity of Science', in W. J. Gonzalez (ed.), *Creativity, Innovation, and Complexity in Science* (A Coruña: Netbiblo, 2013), pp. 19–20.

57. On sustainable development see I. Niiniluoto, 'Nature, Man, and Technology – Remarks on Sustainable Development', in L. Heininen (ed.), *The Changing Circumpolar North: Opportunities for Academic Development* (Rovaniemi: Arctic Centre Publications 6, 1994), pp. 73–87.

58. On the characterization of 'biotechnology' see B. Gremmen, 'Biotechnology: Plants and Animals', in J. K. B. Olsen, S. A. Pedersen and V. F. Hendricks (eds), *A Companion to the Philosophy of Technology*, Blackwell Companions to Philosophy (Chichester: Wiley-Blackwell, 2009), p. 402.

59. K. Lee, 'Biology and Technology', in Olsen, Pedersen and Hendricks (eds), *A Companion to the Philosophy of Technology*, p. 102.

60. See W. J. Gonzalez, 'The Philosophical Approach to Science, Technology and Society', in W. J. Gonzalez (ed.), *Science, Technology and Society: A Philosophical Perspective* (A Coruña: Netbiblo, 2005), pp. 3–49.

61. See I. Niiniluoto, 'Should Technological Imperatives Be Obeyed?', *International Studies in the Philosophy of Science*, 4 (1990), pp. 181–7.

9 Buzzoni, Teleology and Mechanism in Biology

1. J. Lennox, 'Darwin was a Teleologist', *Biology and Philosophy*, 8 (1993), pp. 409–22.

2. J. Jacob, 'Teleology and Reduction in Biology', *Biology and Philosophy*, 1 (1986), pp. 389–99, on p. 398; A. Woodfield, *Teleology* (Cambridge: Cambridge University Press, 1976); M. Bedau, 'Where's the Good in Teleology?', *Philosophy and Phenomenological Research*, 52 (1992), pp. 781–806; B. Maund, 'Proper Functions and Aristotelian Func-

tions in Biology', *Studies in History and Philosophy of Biological and Biomedical Sciences*, 31 (2000), pp. 155–78.

3. E. Nagel, *The Structure of Science* (London and New York: Harcourt-Brace, 1961); C. G. Hempel, 'The Logic of Functional Analysis', in C. G. Hempel, *Aspects of Scientific Explanation and Other Essays* (New York: Free Press, 1965), pp. 297–330.

4. See, above all, L. Wright, 'Functions', *Philosophical Review*, 82 (1973), pp. 139–68 and L. Wright, *Teleological Explanation: An Etiological Analysis of Goals and Functions,* (Berkeley and Los Angeles, CA: University of California Press, 1976); W. C. Wimsatt, 'Teleology and the Logical Structure of Function Statements', *Studies in the History and Philosophy of Science*, 3 (1972), pp. 1–80; C. Boorse, 'Wright on Functions', *Philosophical Review*, 85 (1976), pp. 70–86; E. Berent, 'Function Attributions and Functional Explanations', *Philosophy of Science*, 46 (1979), pp. 343–65; R. G. Millikan, *Language, Thought and Other Biological Categories: New Foundations for Realism* (Cambridge, MA: MIT Press, 1984); R. G. Millikan, 'In Defense of Proper Functions', *Philosophy of Science*, 56 (1989), pp. 288–302; K. Neander, 'What Does Natural Selection Explain? Correction to Sober', *Philosophy of Science*, 55 (1988), pp. 422–6; K. Neander, 'Functions as Selected Effects: The Conceptual Analyst's Defense', *Philosophy of Science*, 58 (1991), pp. 168–84; C. Allen and M. Bekoff, 'Biological Function, Adaptation, and Natural Design', *Philosophy of Science*, 62 (1995), pp. 609–22; M. Mossio, C. Saborido and A. Moreno, 'An Organizational Account of Biological Functions', *British Journal for Philosophy of Science*, 60 (2009), pp. 813–41; E. Kingma, 'Paracetamol, Poison, and Polio: Why Boorse's Account of Function Fails to Distinguish Health and Disease', *British Journal for Philosophy of Science*, 61 (2010), pp. 241–64. The same is also true of Cummin's causal role theory of proper functions (see for example R. Cummins, 'Functional Analysis', *Journal of Philosophy*, 72 (1975), pp. 741–65).

5. A. Weismann, *Vorträge über Deszendenztheorie*, 2 vols, 2nd edn (1902; Jena, Fischer, 1904), vol. 1, p. 47.

6. R. Dawkins, *The Blind Watchmaker: Why the Evidence of Evolution Reveals a Universe without Design* (1986; New York and London: Norton & Company, 1996), p. 5.

7. Dawkins, *The Blind Watchmaker*, p. 1.

8. R. A. Jr Skipper and R. L. Millstein, 'Thinking about Evolutionary Mechanisms: Natural Selection', *Studies in History and Philosophy of Biological and Biomedical Sciences*, 36 (2005), pp. 327–47.

9. C. F. Craver and L. Darden, 'Introduction', *Studies in History and Philosophy of Biological and Biomedical Sciences*, 36 (2005), pp. 233–44, on p. 235.

10. See, in particular, E. Mayr, 'Cause and Effect in Biology: Kinds of Causes, Predictability, and Teleology are Viewed by a Practicing Biologist', *Science*, 134 (1961), pp. 1501–6; E. Mayr, *Toward a New Philosophy of Biology* (Cambridge, MA: Harvard University Press, 1988); F. A. Ayala, 'Teleological Explanations in Evolutionary Biology', *Philosophy of Science,* 37 (1970), pp. 1–15; D. B. Resnik, 'Functional Language and Biological Discovery', *Journal for General Philosophy of Science*, 26 (1995), pp. 119–34; P. S. Agutter and D. N. Wheatley, 'Foundations of Biology: On the Problem of "Purpose" in Biology in Relation to Our Acceptance of the Darwinian Theory of Natural Selection', *Foundations of Science*, 4:1 (1999), pp. 3–23; M. Quarfood, 'Kant on Biological Teleology: Towards Two-Level Interpretation', *Studies in History and Philosophy of Biological and Biomedical Sciences*, 37 (2006), pp. 735–47. Even though they differ greatly from each other, all these works have more or less important points of contact with the viewpoint developed in this chapter.

11. In what follows I shall use the word 'methodical' to denote the methods which can be used as means to a cognitive end, such as Descartes's 'methodical doubt' or 'methodical rules' can exemplify; in my opinion, it is advisable to use the term 'methodological' to denote the discourse on the methods, in a sense which applies to, say, a textbook on methods.

12. Not only 'the new mechanistic philosophy', but also the German methodical or cultural constructivism is relevant to the claim that biology is in principle an experimental science (see, above all, P. Janich, 'Experiment in der Biologie', *Theory in Biosciences*, 116:1 (1997), Repr. in P. Janich, *Kultur und Methode. Philosophie in einer wissenschaftlich geprägten Welt* (Frankfurt: Suhrkamp, 2006), pp. 330–66.; R. Lange, *Experimentalwissenschaft Biologie. Methodische Grundlagen und Probleme einer technischen Wissenschaft vom Lebendigen* (Würzburg: Königshausen+Neumann, 1999).

13. P. Machamer, L. Darden and C. F. Craver, 'Thinking about Mechanisms', *Philosophy of Science*, 67 (2000), pp. 1–25, on p. 3.

14. S. Glennan, 'Rethinking Mechanistic Explanation', *Philosophy of Science*, 69 (2002), pp. 342–53, on p. 344.

15. W. Bechtel, *Discovering Cell Mechanisms: The Creation of Modern Cell Biology* (Cambridge: Cambridge University Press, 2006), p. 26.

16. On the epistemological and ontological differences between Glennan's and Machamer, Darden and Craver's concept of mechanism, see, above all, J. G. Tabery, 'Synthesizing Activities and Interactions in the Concept of a Mechanism', *Philosophy of Science*, 71 (2004), pp. 1–15.

17. R. Harré, *The Principles of Scientific Thinking* (London: MacMillan, 1970), p. 109. In Harré, we find many of the key notions of the more recent debate about mechanisms: the 'generative' character of the causal relation, the 'interactions' between the parts of a mechanism, the causal 'responsibility' of mechanisms (see e.g. W. Bechtel, *Discovering Cell Mechanisms: The Creation of Modern Cell Biology* (Cambridge: Cambridge University Press, 2006): 'The orchestrated functioning of the mechanism is responsible for one or more phenomena', p. 26), and so on.

18. See for example J. L. Mackie, *The Cement of the Universe: A Study of Causation* (Oxford: Clarendon Press, 1974), pp. 228–99, and W. Salmon, *Scientific Explanation and the Causal Structure of the World* (Princeton, NJ: Princeton University Press, 1984), p. 275. Mackie's and Salmon's theory of causality are extensively discussed by the main exponents of the new mechanistic approach and by their critics: see e.g. respectively S. Glennan, 'Ephemeral Mechanisms and Historical Explanation', *Erkenntnis*, 72 (2010), pp. 251–66 and S. Psillos, 'A Glimpse of the Secret Connexion: Harmonizing Mechanisms with Counterfactuals', *Perspectives on Science*, 12 (2004), pp. 288–319.

19. W. Bechtel, 'The Downs and Ups of Mechanistic Research: Circadian Rhythm Research as an Exemplar', *Erkenntnis*, 73 (2010), pp. 313–28.

20. See D. J. Nicholson, 'The Concept of Mechanism in Biology', *Studies in History and Philosophy of Biological and Biomedical Sciences*, 43 (2012), pp. 152–63, on pp. 152–3; see also P. Machamer, L. Darden and C. F. Craver, 'Thinking about Mechanisms', *Philosophy of Science*, 67 (2000), pp. 1–25, on p. 1, who write: 'Thinking in terms of mechanisms provides a new framework for addressing many traditional philosophical issues: causality, laws, explanation, reduction, and scientific change'. And Glennan has gone so far as to argue that also the events of the human history, notwithstanding their 'ephemeral nature', are processes that deserve to be called mechanisms (S. Glennan, 'Ephemeral Mechanisms and Historical Explanation', *Erkenntnis*, 72 (2010), pp. 251–66).

21. See for example J. Woodward, 'What Is a Mechanism? A Counterfactual Account', *Philosophy of Science*, 69 (2002), pp. S366–S377; J. Woodward, 'Mechanisms Revisited', *Philosophy of Science*, 183 (2011), pp. 409–27; S. Psillos, 'A Glimpse of the Secret Connexion: Harmonizing Mechanisms with Counterfactuals', *Perspectives on Science*, 12 (2004), pp. 288–319; J. G. Tabery, 'Synthesizing Activities and Interactions in the Concept of a Mechanism', *Philosophy of Science*, 71 (2004), pp. 1–15; D. J. Nicholson, 'The Concept of Mechanism in Biology', *Studies in History and Philosophy of Biological and Biomedical Sciences*, 43 (2012), pp. 152–63. The most important partial exception is Braillard, but his conclusions are, after all, too eclectic. On the one side, he concedes that the mechanist approach is in principle consistent with systems biology: 'systems biology is clearly concerned with large and complex mechanisms, but they are still mechanisms'. (P.-A. Braillard, 'Systems Biology and the Mechanistic Framework', *History & Philosophy of the Life Sciences*, 32 (2010), pp. 43–62, on p. 45). On the other side, he claims that 'design explanations' are in principle different from mechanistic ones: 'the two major differences between design explanations and mechanistic explanations need to be clarified. First, design explanations are non-causal and are concerned with constraints, not with temporal processes. The second difference concerns their generality, since design explanations point to general principles, something not common in molecular biology'. (P.-A. Braillard, 'Systems Biology and the Mechanistic Framework', *History & Philosophy of the Life Sciences*, 32 (2010), pp. 43–62, on p. 50). In any event, Braillard deserves credit for having introduced, without being aware of it, some of the well-known ideas of Gould and Lewontin (S. J. Gould and R. C. Lewontin, 'The Spandrels of San Marco and the Panglossian Paradigm: A Critique of the Adaptationist Program', *Proceedings of the Royal Society of London*, B 205 (1979), pp. 581–9) in the debate about mechanical explanations: compare for example Gould and Lewontin's key expressions such as 'structure' and 'architectural constraints' with Braillard's 'structural properties' (p. 55) and 'architectural principles' (p. 50), or with his claim according to which 'design explanations show how functions constrain structures' (p. 55).

22. It is no accident, for example, that Glennan (S. Glennan, 'Ephemeral Mechanisms and Historical Explanation', *Erkenntnis*, 72 (2010), pp. 251–66), who exploits the mechanism concept to understand explanation in history and in the human sciences, gives no attention to the analysis of such concepts as teleology, purpose, aim or end (he does not even mention them).

23. See Craver and Bechtel (C. F. Craver and W. Bechtel, 'Top-Down Causation without Top-Down Causes', *Biology and Philosophy*, 22 (2007), 547–63, on p. 547: 'Many philosophers ... appeal to top-down causes in their explanations. Such appeals evoke concerns that the notion of top-down causation is incoherent or that it involves spooky forces exerted by wholes upon their components ... Appeal to top-down causation seems spooky or incoherent when it cannot be explicated in terms of mechanistically mediated effects'.

24. M. Polanyi, *Personal Knowledge: Towards a Post-Critical Philosophy* (1958; London: Routledge & Kegan Paul, 1962) (quotations are from the 1962 revised edition), p. 378 (see also p. 344). According to Polanyi, however, the difference between a machine and a living being consists in what he calls 'the inventive powers of animal life': 'while the animal's machinery embodies fixed operational principles, this machinery would be impelled, guided and readapted by the animal's unspecifiable inventive urge'. On this point, I disagree with Polany because this claim interprets teleology as a real property of living beings without attributing any methodical value to it.

25. D. J. Nicholson, 'The Concept of Mechanism in Biology', *Studies in History and Philosophy of Biological and Biomedical Sciences*, 43 (2012), 152–63, on p. 158).

26. P. Thagard, 'Explaining Disease: Causes, Correlations, and Mechanisms', *Minds and Machines*, 8 (1998), pp. 61–78, on p. 66.

27. S. Glennan, 'Mechanisms and the Nature of Causation', *Erkenntnis*, 44 (1996), pp. 49–71, on p. 51.

28. P. Machamer, L. Darden and C. F. Craver 'Thinking about Mechanisms', *Philosophy of Science*, 67 (2000), pp. 1–25, on p. 17.

29. If we exclude intuitions and isolated remarks found in many authors, Dingler (H. Dingler, *Die Methode der Physik* (Munich: Reinhardt, 1938), cap. II, part IV, § 3) and Collingwood (R. G. Collingwood, *An Essay on Metaphysics* (Oxford: Clarendon Press, 1940) were probably the first to methodically develop the idea that the concepts of 'cause' and 'causal explanation' depend on our capacity to intervene practically in reality. They argued that if we relied only on passively received sensations reporting successions of events, we would be unable to grasp causal relations and formulate causal explanations. This idea was then taken up by Gasking (D. Gasking, 'Causation and Recipes', *Mind*, 64 (1955), pp. 479–87), Gorovitz (S. Gorovitz, 'Causal Judgements and Causal Explanations', *Journal of Philosophy*, 62 (1965), pp. 695–711), von Wright (G. H. von Wright, *Explanation and Understanding* (New York: Cornell University Press, 1971 and G. H. von Wright, 'A Reply to My Critics', in P. A. Schilpp and L. E. Hahn (eds), *The Philosophy of Georg Henrik von Wright* (La Salle, IL: Open Court, 1989), pp. 733–887) and, more recently, by a number of authors such as Price (H. Price, 'Agency and Probabilistic Causality', *British Journal for the Philosophy of Science*, 42 (1991), pp. 157–76; H. Price, 'Agency and Causal Asymmetry', *Mind*, 101 (1992), pp. 501–20; H. Price, 'Causal Perspectivalism', in H. Price and R. Corry (eds), *Physics and the Constitution of Reality: Russell's Republic Revisited* (Oxford: Oxford University Press, 2007), pp. 250–92), Menzies and Price (P. Menzies and H. Price, 'Causation as a Secondary Quality', *British Journal for the Philosophy of Science*, 42 (1993), 187–203, on p. 187), Hausman (D. M. Hausman, *Causal Asymmetries* (Cambridge: Cambridge University Press, 1998)), Keil (G. Keil, *Handeln und Verursachen* (Frankfurt: Klostermann, 2000)), Pearl (J. Pearl, *Causality* (Cambridge: Cambridge University Press, 2000) and Woodward (J. Woodward, *Making Things Happen: A Theory of Causal Explanation* (Oxford: Oxford University Press, 2003); J. Woodward, 'Agency and Interventionist Theories' in H. Bebee, P. Menzies and C. Hitchcock (eds), *The Oxford Handbook of Causation* (Oxford: Oxford University Press, 2009), pp. 234–62; J. Woodward, 'Causation in Biology: Stability, Specificity, and the Choice of Levels of Explanation', *Biology and Philosophy*, 25 (2010), pp. 287–318).

30. M. Buzzoni, 'Causality, Experiment, and Anthropomorphism', in E. Agazzi (ed.), *Representation and Explanation in the Sciences* (Milan: Angeli, 2013), pp. 143–53; M. Buzzoni, *Scienza e tecnica. Teoria ed esperienza nelle scienze della natura* (Roma: Studium, 1995), ch. 2, § 1; M. Buzzoni, Causalité et temporalité du point de vue opérationnel', *Epistemologia: An Italian Journal for the Philosophy of Science*, Special Issue, 14 (2009), *Time in the Different Approaches/Le temps appréhendé à travers différentes disciplines*, pp. 45–58.

31. J. Woodward, *Making Things Happen: A Theory of Causal Explanation* (Oxford: Oxford University Press, 2003), ch. 3.

32. Woodward (Woodward, *Making Things Happen*, p. 120. For the thesis that causality is relative to our theoretical or practical interests, see especially Dingler (H. Dingler, *Die*

Methode der Physik (Munich: Reinhardt, 1938), Gardiner (P. Gardiner, *The Nature of Historical Explanation* (London: Oxford University Press, 1952), Hanson (N. R. Hanson, *Patterns of Discovery*: Cambridge: Cambridge University Press, 1958), Gorovitz (S. Gorovitz, 'Causal Judgements and Causal Explanations', *Journal of Philosophy*, 62 (1965), pp. 695–711, on p. 695), von Wright (G. H. von Wright, *Explanation and Understanding* (New York: Cornell University Press, 1971), p. 93), Putnam (H. Putnam, *Renewing Philosophy* (Harvard, MA: Harvard College, 1992), ch. 5), Buzzoni (M. Buzzoni, *Scienza e tecnica. Teoria ed esperienza nelle scienze della natura* (Roma: Studium, 1995); M. Buzzoni, 'Causality, Experiment, and Anthropomorphism', in E. Agazzi (ed.), *Representation and Explanation in the Sciences* (Milan: Angeli, 2013), pp. 143–53), Price (H. Price, 'Causal Perspectivalism', in H. Price and R. Corry (eds), *Physics and the Constitution of Reality: Russell's Republic Revisited* (Oxford: Oxford University Press, 2007), pp. 279–88). On the necessity of relativizing the notion of mechanism, see Nicholson (D. J. Nicholson, 'The Concept of Mechanism in Biology', *Studies in History and Philosophy of Biological and Biomedical Sciences*, 43 (2012), pp. 152–63.

33. Woodward, *Making Things Happen*, p. 104.

34. Von Wright, *Explanation and Understanding*, p. 117.

35. P. Menzies and H. Price, 'Causation as a Secondary Quality', *British Journal for the Philosophy of Science*, 42 (1993), 187–203, on p. 187.

36. See Kant, CPR: B xiii–xiv.

37. According to Woodward, an agency theory of causality along these lines is excessively anthropomorphic (Woodward, *Making Things Happen*, p. 123). For a detailed discussion and rejection of the Woodward's charge of anthropomorphism, see Buzzoni, 'Causality, Experiment, and Anthropomorphism'. On the contrary, in accordance with the view I have been advocating is Peter Janich's 'fundamental argument' against empiricist epistemology (see P. Janich, 'Was macht experimentelle Resultate empiriehaltig? Die methodisch-kulturalistische Theorie des Experiments', in M. Heidelberger and F. Steinle (eds), *Experimental Essays – Versuche zum Experiment* (Baden-Baden: Nomos, 1998), pp. 100–1). As he noted, drawing on Dingler's philosophy of science, we cannot distinguish between what is and what is not a clock on the basis of a natural law, since broken clocks too work in accordance with what scientists call 'natural laws'. To make a distinction, we need also the indication of a human purpose, which the broken clocks do not satisfy (P. Janich, *Grenzen der Naturwissenschaft* (Munich: Beck, 1992), pp. 194, 205).

38. E. Mayr, *Toward a New Philosophy of Biology* (Cambridge, MA: Harvard University Press, 1988), pp. 62–3.

39. In philosophy of science, it was above all Evandro Agazzi who emphasized this point: see for example E. Agazzi, *Temi e problemi di filosofia della fisica* (Milan: Manfredi, 1969).

40. For more details on this point, see M. Buzzoni, *Thought Experiment in the Natural Sciences* (Würzburg: Königshausen+Neumann, 2008).

41. M. D. Laubichler and G. P. Wagner, 'Organism and Character Decomposition: Steps towards an Integrative Theory of Biology', *Philosophy of Science*, 67 (2000), pp. S289–S300; J. W. Pepper and M. D. Herron, 'Does Biology Need an Organism Concept?', *Biol. Review*, 83 (2008), pp. 621–7.

42. On the dispensability or centrality of the concept of organism in biology, see also the opposite views maintained by B. D. Dyer, 'Symbiosis and Organismal Boundaries', *American Zoologist*, 29 (1989), pp. 1085–93, and C. N. El-Hani and C. Emmeche 'On Some Theoretical Grounds for an Organism-Centered Biology: Property Emergence,

Supervenience, and Downward Causation', *Theory in Bioscience*, 119 (2000), pp. 234–75.

43. S. Y. Lee, A. Lee, J. Y. Chen and R. MacKinnon, 'Structure of the KvAP Voltage-Dependent K1 Channel and its Dependence on the Lipid Membrane', *Proc. National Acad. Sci. USA*, 102 (2005), pp. 15441–6; D. Schmidt, J. Qiu-Xing and R. MacKinnon, 'Phospholipids and the Origin of Cationic Gating Charges in Voltage Sensors', *Nature*, 444/7 (2006), pp. 775–9; A. Alessandrini, P. Gavazzo, C. Picco and P. Facci, 'Voltage-Induced Morphological Modifications in Oocyte Membranes Containing Exogenous K1 Channels Studied by Electrochemical Scanning Force Microscopy', *Microscopy Research and Technique*, 71 (2008), pp. 274–8.

44. Y. Li-Smerin and K. J. Swartz, 'Helical Structure of the COOH Terminus of S3 and its Contribution to the Gating Modifier Toxin Receptor in Voltage-Gated Ion Channels', *Journal of General Physiology*, 117 (2001), pp. 205–17, on p. 117; J. M. Wang, S. H. Roh, K. Sunghwan, C. W. Lee, I. K. Jae and K. J. Swartz, 'Molecular Surface of Tarantula Toxins Interacting with Voltage Sensors in Kv Channels', *Journal for General Physiology*, 123 (2004), pp. 455–67.

45. K. Lorenz, 'Kants Lehre vom Apriorischen im Lichte gegenwärtiger Biologie', *Blätter für deutsche Philosophie*, 15 (1941/2), pp. 94–125, on pp. 98–9.

10 Diéguez, Scientific Understanding and the Explanatory Use of False Models

1. Research for this work has been supported by the research projects FFI2012–37354 (Spanish Government), HUM–0264 and HUM–7248 (Junta de Andalucía). I'm also very grateful to João Queiroz for useful discussions.

2. J. Bransen, '*Verstehen* and *Erklären*, Philosophy of', in N. J. Smelser and P. B. Baltes (eds), *International Encyclopedia of the Social and Behavioral Sciences* (Oxford: Elsevier Science Ltd, 2001), pp. 16165–70.

3. W. C. Salmon, *Causality and Explanation* (Oxford: Oxford University Press, 1998).

4. P. Kitcher, 'Explanatory Unification', *Philosophy of Science*, 48 (1981), pp. 507–531, on p. 508.

5. See E. McMullin, 'Structural Explanation', *American Philosophical Quarterly*, 15 (1978), pp. 139–47; N. Cartwright, *How the Laws of Physics Lie* (Oxford: Clarendon Press, 1983) ch. 8; E. Sober, 'Equilibrium Explanation', *Philosophical Studies*, 43 (1983), pp. 201–10; W. C. Wimsatt, 'False Models as Means to Truer Theories', in M. H. Nitecki and A. Hoffman (eds), *Neutral Models in Biology* (New York: Oxford University Press, 1987), pp. 23–55; P. K. Machamer, L. Darden and C. F. Craver, 'Thinking about Mechanisms', *Philosophy of Science*, 57 (2000), pp. 1–25; M. Elgin and E. Sober, 'Cartwright on Explanation and Idealization', *Erkenntnis*, 57 (2002), pp. 441–50; A. Plutynski, 'Explanation in Classical Population Genetics', *Philosophy of Science*, 71 (2004), pp. 1201–14; W. Bechtel and A. Abrahamsen, 'Explanation: A Mechanist Alternative', *Stud. Hist. Phil. Biol. & Biomed. Sci.*, 36 (2005), pp. 421–41; J. Odenbaugh, 'Idealized, Inaccurate but Successful: A Pragmatic Approach to Evaluating Models in Theoretical Ecology', *Biology and Philosophy*, 20 (2005), pp. 231–55; C. Marchionni, 'Contrastive Explanation and Unrealistic Models: The Case of the New Economic Geography', *Journal of Economic Methodology*, 13:4 (2006), pp. 425–46; C. F. Craver, 'When Mechanistic Models Explain', *Synthese*, 153 (2006), pp. 355–76;

L. Darden, 'Thinking Again about Biological Mechanisms', *Philosophy of Science*, 75 (2008), pp. 958–69; F. Hindriks, 'False Models as Explanatory Engines', *Philosophy of the Social Sciences*, 38:3 (2008), pp. 334–60; F. Hindriks, 'Explanation, Understanding, and Unrealistic Models', *Studies in History and Philosophy of Science*, 44 (2013), pp. 523–31; M. Weisberg, 'Who is a Modeler?', *British Journal for the Philosophy of Science*, 58 (2007), pp. 207–33; A. Bokulich, 'How Scientific Models Can Explain', *Synthese*, 180:1 (2011), pp. 33–45; A. G. Kennedy, 'A Non-Representationalist View of Model Explanation', *Studies in History and Philosophy of Science*, 43:2 (2012), pp. 326–32.

6.   D. Portides, 'Models', in S. Psillos and M. Curd (eds), *The Routledge Companion to Philosophy of Science* (London: Routledge, 2008), pp. 385–95, on p. 385.

7.   See M. Morrison, 'Models as Autonomous Agents', in M. Morgan and M. Morrison (eds), *Models as Mediators. Perspectives on Natural and Social Science* (Cambridge: Cambridge University Press, 1999), pp. 38–65 and M. Weisberg, 'Three Kinds of Idealization', *Journal of Philosophy*, 104:12 (2007), pp. 639–59.

8.   For example, T. Knuuttila and M. Merz, 'Understanding by Modeling: An Objectivist Approach', in H. W. de Regt, S. Leonelli and K. Eigner (eds), *Scientific Understanding: Philosophical Perspectives* 2009, pp. 146–65; and S. Leonelli, 'Understanding in Biology', in H.W. de Regt, S. Lionelli and K. Eigner (eds), *Scientific Understanding: Philosphical Perspectives* 2009, pp. 189–209.

9.   For a discussion, see P. Lipton, 'Understanding without Explanation', in H. W. de Regt, S. Leonelli and K. Eigner (eds), *Scientific Understanding: Philosophical Perspectives*, 2009, pp. 43–63; V. Gijsbers, 'Understanding, Explanation, and Unification', *Studies in History and Philosophy of Science*, 44 (2013), pp. 516–22; M. Strevens, 'No Understanding without Explanation', *Studies in History and Philosophy of Science*, 44 (2013), pp. 510–15.

10.   J. Kvanvig, *The Value of Knowledge and the Pursuit of Understanding* (Cambridge: Cambridge University Press, 2003), pp. 190–1.

11.   H.W. de Regt, S. Leonelli and K. Eigner (eds), *Scientific Understanding: Philosophical Perspectives* (Pittsburgh, PA: University of Pittsburgh Press, 2009), p. 2.

12.   M. Friedman, 'Explanation and Scientific Understanding', *Journal of Philosophy*, 72:1 (1974), pp. 5–19, on p. 15.

13.   G. Schurz and K. Lambert, 'Outline of a Theory of Scientific Understanding', *Synthese*, 101 (1994), pp. 65–120.

14.   H. W. de Regt and D. Dieks, 'A Contextual Approach to Scientific Understanding', *Synthese*, 144 (2005), pp. 137–170.

15.   J. Kuorikoski, 'Simulation and the Sense of Understanding', in P. Humphreys and C. Imbert (eds), *Models, Simulations, and Representations* (New York: Routledge, 2011), pp. 168–87.

16.   M. Strevens, 'No Understanding without Explanation', *Studies in History and Philosophy of Science*, 44 (2013), pp. 510–15.

17.   V. Gijsbers, 'Understanding, Explanation, and Unification', *Studies in History and Philosophy of Science*, 44 (2013), pp. 516–22.

18.   K. Khalifa and M. Gadomski, 'Understanding as Explanatory Knowledge: The Case of Bjorken Scaling', *Studies in History and Philosophy of Science*, 44 (2013), pp. 384–92.

19.   D.A. Wilkenfeld, 'Understanding as Representation Manipulability', *Synthese*, 190 (2013), pp. 997–1016.

20.   C.Z. Elgin, 'Is Understanding Factive?', in A. Haddock, A. Millar, and D. Pritchard (eds), *Epistemic Value* (Oxford: Oxford University Press, 2009), p. 327.

21. P. Godfrey-Smith, 'The Strategy of Model-based Science', *Biology and Philosophy*, 21 (2006), pp. 725–40.

22. J. Kvanvig, *The Value of Knowledge and the Pursuit of Understanding* (Cambridge: Cambridge University Press, 2003), ch 8.

23. R. Giere, *Science without Laws* (Chicago, IL: University of Chicago Press, 1999).

24. See P. Godfrey-Smith, 'Models and Fictions in Science', *Philosophical Studies*, 143 (2009), pp. 101–16; R. Frigg, 'Models and Fictions', *Synthese*, 172 (2010), pp. 251–68; G. Contessa, 'Scientific Models and Fictional Objects', *Synthese*, 172 (2010), pp. 215–29.

25. See P. Kitcher, *Science, Truth, and Democracy* (Oxford: Oxford University Press, 2001), ch. 5; and, for a different opinion, S. Sismondo and N. Chrisman, 'Deflationary Metaphysics and the Natures of Maps', *Philosophy of Science*, 68 (2001), pp. S38–S49.

26. For a discussion and some important clarifications, see A. Chakravartty, 'Truth and Representation in Science: Two Inspiration from Art', in R. Frigg and M. C. Hunter (eds), *Beyond Mimesis and Conventions. Representation in Art and Science* (Dordrecht: Springer, 2010), pp. 33–50 and U. Mäki, 'Models and the Locus of their Truth', *Synthese*, 180 (2011), pp. 47–63.

27. M. Elgin 'Understanding's Tethers', in C. Jäger and W. Löffler (eds), *Epistemology: Contexts, Values, Disagreement* (Frankfurt: Ontos Verlag, 2012), pp. 131–45.

28. F. Hindriks, 'False Models as Explanatory Engines', *Philosophy of the Social Sciences*, 38:3 (2008), pp. 334–60.

29. M. Morrison, 'Understanding in Physics and Biology', in H. W. de Regt, S. Leonelli and K. Eigner (eds), *Scientific Understanding. Philosophical Perspectives* (2009), pp. 123–45.

30. A. G. Kennedy, 'A Non-Representationalist View of Model Explanation', *Studies in History and Philosophy of Science*, 43:2 (2012), pp. 326–32.

31. The three first items in this typology can be seen somewhat like a simplification of Wimsatt's classification of false models(W. C. Wimsatt, 'False Models as Means to Truer Theories', in M. H. Nitecki and A. Hoffman (eds), *Neutral Models in Biology* (New York: Oxford University Press, 1987), pp. 23–55).

32. See J. Rodríguez, *Ecología* (Madrid: Pirámide, 1999), pp. 274–87.

33. Wimsatt, 'False Models as Means to Truer Theories'.

34. Wimsatt, 'False Models as Means to Truer Theories', p. 27.

35. Wimsatt, 'False Models as Means to Truer Theories', p. 28.

36. E. Sober, *The Nature of Selection* (Chicago, IL: University of Chicago Press, 1984), p. 23.

37. M. Ridley, *Evolution*, 2nd edn (Cambridge, MA: Blackwell, 1996), pp. 307–12.

38. M. Morrison, 'Fictions, Representations, and Reality', in M. Suárez (ed.), *Fictions in Science* (New York: Routledge, 2009), pp. 110–35.

39. C. Z. Elgin, 'Telling Instances', in R. Frigg and M. C. Hunter (eds), *Beyond Mimesis and Conventions: Representation in Art and Science* (Dordrecht: Springer, 2010), pp. 1–17.

40. A. Toon, 'Models as Make-Believe', in Frigg and Hunter (eds), *Beyond Mimesis and Conventions*, pp. 71–96.

41. M. Weisberg, 'Who Is a Modeler', *British Journal for the Philosophy of Science*, 58 (2007), pp. 207–33, on p. 223.

42. A. Diéguez, 'When Do Models Provide Genuine Understanding, and Why Does It Matter?', *History and Philosophy of the Life Sciences*, 35:4 (2013), pp. 599–620.

43. L. D. Hurst, 'Why are There Only Two Sexes?', *Proc. R. Soc. Lond. B.*, 263 (1996), pp. 415–22.

44. T. L. Czárán and R. F. Hoekstra, 'Evolution of Sexual Asymmetry', *BMC Evolutionary Biology*, 4:34 (2004), pp. 1–12.

45. Czárán and Hoekstra, 'Evolution of Sexual Asymmetry', p. 7.

46. M. B. Bonsall, 'The Evolution of Anisogamy: The Adaptive Significance of Damage, Repair and Mortality', *Journal of Theoretical Biology*, 238 (2006), pp. 198–210.

47. The ciliated protozoan *Tetrahymena thermophile* is one of these rare organisms. It has seven mating types. Some eusocial insects, like ants of the genus *Pogonomyrmex*, have a three-sex reproductive system.

48. J. Kvanvig, 'Response to Critics', in A. Haddock, A. Millar and D. Pritchard (eds), *Epistemic Values* (Oxford: Oxford University Press, 2009), pp. 339–351, on p. 342.

49. C. Z. Elgin, 'True Enough', *Philosophical Issues*, 14:1 (2004), pp. 113–31; C. Z. Elgin, 'Telling Instances', in R. Frigg and M. C. Hunter (eds), *Beyond Mimesis and Conventions: Representation in Art and Science* (Dordrecht: Springer, 2010), pp. 1–17.

50. Elgin, 'True Enough', pp. 126–7.

51. Paul Humphrey sees here a limit for the explanatory use of this kind of models. They could provide understanding, but not explanation. He writes: 'here, then, is perhaps where one part of the boundary between explanation and understanding lies. Although it can enhance our scientific understanding to explore models that violate the laws of our universe, such models cannot be used in explanations. A well-known example involves the conditions under which life can emerge in the universe. The 'how possibly?' questions investigated in the neighbourhood of anthropic principles add to our understanding of how life might have emerged if the laws had been different, but answers to them cannot explain life as it arose in our universe' (Humphrey, 'Invariance, Explanation, and Understanding', pp. 42–3). The three examples I mentioned show, however, that Humphreys' scruples are exaggerated and that these models can have sometimes an explanatory function.

52. Wimsatt, 'False Models as Means to Truer Theories', p. 30.

53. C. Swoyer, 'Structural Representation and Surrogative Reasoning', *Synthese*, 87 (1991), pp. 449–508; M. Suárez, 'An Inferential Conception of Scientific Representation', *Philosophy of Science*, 71 (2004), pp. 767–79; G. Contessa, 'Scientific Models and Fictional Objects', *Synthese*, 172 (2010), pp. 215–29.

54. But see I. Niiniluoto, *Critical Scientific Realism* (Oxford: Oxford University Press, 1999), p. 192.

55. See F. M. Harman, *Energy, Force, and Matter: The Conceptual Development of Nineteenth-Century Physics* (Cambridge: Cambridge University Press, 1982), ch. 4; although see A. Chalmers, 'The Heuristic Role of Maxwell's Mechanical Model of Electromagnetic Phenomena', *Studies in History and Philosophy of Science*, 17:4 (1986), pp. 415–27 for a contrary point of view.

56. M. Morrison, 'Approximating the Real: The Role of Idealizations in Physical Theory', in M. R. Jones and N. Cartwright (eds), *Idealization XII: Correcting the Model. Idealization and Abstraction in the Sciences (Poznań Studies in the Philosophy of the Sciences and the Humanities, vol. 86)* (Amsterdam and New York: Rodopi, 2005), p. 170.

57. C.G. Hempel, *Aspects of Scientific Explanation and Other Essays* (New York: Free Press, 1965).

58. J. D. Trout, 'Scientific Explanation and the Sense of Understanding', *Philosophy of Science*, 69 (2002), pp. 212–33.

59. M. Friedman, 'Explanation and Scientific Understanding', *Journal of Philosophy*, 72:1 (1974), pp. 5–19, on p. 8. For an insightful reply to Trout's paper, see H. W. de Regt, 'Discussion Note: Making Sense of Understanding', *Philosophy of Science*, 71 (2004), pp. 98–109.

60. C.C. Elgin, 'Review of Henk W. de Regt, Sabina Leonelli, and Kai Eigner (eds), *Scientific Understanding: Philosophical Perspectives* (University of Pittsburgh Press, 2010) Notre Dame Philosophical Reviews, at http://ndpr.nd.edu/news/24266-scientific-understanding-philosophical-perspectives/ [accessed 13 January 2014].
61. A. Diéguez, 'When Do Models Provide Genuine Understanding, and Why Does It matter?', *History and Philosophy of the Life Sciences*, 35:4 (2013), pp. 599–620.
62. M. Weisberg, 'Who is a Modeler?', *British Journal for the Philosophy of Science*, 58 (2007), pp. 207–33.

11 Bertolaso, Di Stefano, Ghilardi and Marcos, Bio-Techno-Logos and Scientific Practice

1. On this subject see A. Marcos, *Filosofia dell'agire scientifico, nuove dimensioni* (Milan: Academia Universa Press, 2010).
2. This reasonable attitude is the scientific correspondent of the Aristotelian virtue of *prudence* (φρόνησις). See A. Marcos, *Postmodern Aristotle* (Newcastle: Cambridge Scholars Publishing, 2012).
3. On the amplitude of Logos in science and biology in particular see the recent A. Nordmann, 'Philosophy of Synthetic Biology: Contested Images of Knowledge Production', B. Giese, A. Von Gleich, C. Pade and H. Wigger (eds), *Synthetic Biology: Character and Impact* (Berlin: Springer, in press).
4. P. Duhem, *La Théorie Physique*, 2nd edn (París: Marcel Rivière, 1914), pp. 390–1. On the relationship between creativity and technology see W. J. Gonzalez, 'The Roles of Scientific Creativity and Technological Innovation in the Context of Complexity of Science', in W. J. Gonzalez (ed.), *Creativity, Innovation, and Complexity in Science* (A Coruña: Netbiblo, 2013), pp. 11–40.
5. I refer here to the experience in a broad sense, in the sense thinkers as Aristotle and contemporary pragmatists give to this notion. It's the whole experience conceived as the time we have lived and learned from, as a reflective experience. It is not, of course, the empiricist or positivist notion of experience.
6. We are following here G. Holton, 'La imaginación en la ciencia', in L. Preta (ed) *Imágenes y metáforas en ciencia* (Madrid: Alianza, 1993), pp. 29–58.
7. In 1609 Galileo made the first realistic renderings of the Moon. These drawings are now housed at the Biblioteca Nazionale Centrale in Florence. In 1610 he included some Moon drawings in his *Siderius Nuntius*, published in Venice. A facsimile edition is available at http://www.rarebookroom.org/Control/galsid/index.html, with drawings of the moon surface at pp. 11–14.
8. On Thomas Harriot, see J, W. Shirley (ed.), *A Sourcebook for the Study of Thomas Harriot* (New York: Arno Press, 1981). His Moon drawings are available at http://galileo.rice.edu/sci/harriot_moon.html.
9. Cf. M. Bertolaso, *Il cancro come questione. Modelli interpretativi e presupposti epistemologici* (Milan: Franco Angeli, 2012).
10. N. Russell, *Communicating Science* (Cambridge: Cambridge University Press, 2010); A. Marcos and J. Chillón, 'Para una comunicación crítica de la ciencia', *Artefactos*, 3:1 (2010), pp. 81–108. For updated information and current theoretical debates on the communication of science see the journal Science Communication (Sage). For updated information and current theoretical debates on science teaching see the journal Science & Education (Springer).

Index

For Product Safety Concerns and Information please contact our EU
representative GPSR@taylorandfrancis.com
Taylor & Francis Verlag GmbH, Kaufingerstraße 24, 80331 München, Germany

www.ingramcontent.com/pod-product-compliance
Lightning Source LLC
Chambersburg PA
CBHW060401220326
41598CB00023B/2988

* 9 781138 706422 *